# Rethinking Mathematics

# RETHINKING
# MATHEMATICS

## Teaching Social Justice by the Numbers
## Second Edition

Edited by
Eric (Rico) Gutstein
and Bob Peterson

A RETHINKING SCHOOLS PUBLICATION

*Rethinking Mathematics*
*Teaching Social Justice by the Numbers*
*Second Edition*
Edited by Eric (Rico) Gutstein and Bob Peterson

Dedicated to
**Claudia Zaslavsky**
1917–2006
Educator, Multiculturalist, Anti-Racist, Ethnomathematician

Rethinking Schools, Ltd., is a nonprofit educational publisher of books, booklets, and a quarterly journal on school reform, with a focus on issues of equity and social justice. To request additional copies of this book or a catalog of other publications, or to subscribe to the quarterly journal *Rethinking Schools,* contact:

Rethinking Schools
1001 East Keefe Avenue
Milwaukee, Wisconsin 53212 USA
800-669-4192
www.rethinkingschools.org

*Rethinking Mathematics: Teaching Social Justice by the Numbers*
Second Edition
© 2013, Rethinking Schools, Ltd.

Cover design: Joanna Dupuis (1st ed.), Nancy Zucker, (2nd ed.)
Cover artwork by David McLimans
Page design and layout: Joanna Dupuis (1st ed.), Kate Hawley (2nd ed.)
Proofreading: Jennifer Morales and Lawrence Sanfilippo

Production Editors: Bill Bigelow and Michael Trokan

Cartoons on pages 64 and 258 are by Fred Wright/UE NEWS. Used with the permission of UE NEWS/United Electrical, Radio and Machine Workers of America (UE).

ISBN: 978-0-942961-55-3 Library of Congress Cataloging-in-Publication Data applied for.

# Contents

## PART THREE: INFUSING SOCIAL JUSTICE MATH INTO OTHER CURRICULAR AREAS

## PART FOUR: RESOURCES FOR RETHINKING MATHEMATICS

# Acknowledgments

Putting together the second edition of *Rethinking Mathematics* would not have been possible if it had not been for the support and work of many individuals. We would like to thank all those math teachers and authors we have worked with to include their writings and ideas in this book. We have learned from their teaching and writing as we hope the readers of this book will as well.

We would like to thank, in particular, Bill Bigelow, Rethinking Schools curriculum editor, who worked closely with us giving advice and feedback on the book as a whole and many specific articles.

Rico Gutstein offers special thanks to the students and communities of Chicago's Social Justice High School (Sojo) and Rivera Elementary School, and to Pauline Lipman for her ever-critical support and analysis.

Bob Peterson thanks the 5th-grade students he has taught at La Escuela Fratney, in Milwaukee, many of whom have provided the inspiration and challenges that have helped him become the social justice teacher that he is. He also thanks the entire Fratney community and dedicated staff, who continue to inspire him. Finally, he thanks Barbara Miner for her support and feedback.

*Rethinking Mathematics,* like all Rethinking Schools books, was supported by the broader Rethinking Schools community, especially the editors and editorial associates, Wayne Au, Terry Burant, Linda Christensen, Helen Gym, David Levine, Stan Karp, Larry Miller, Kelley Dawson Salas, Melissa Bollow Tempel, Rita Tenorio, Stephanie Walters, Dyan Watson, and Kathy Xiong; and the Rethinking Schools staff, past and present: Susan Bates, Catherine Capellaro, Kris Collett, Tegan Dowling, Leon Lynn, Jody Sokolower, and Mike Trokan. And special thanks to Rethinking Schools volunteer Sandy Shedivy for additional research and proofreading.

# Preface

A lot has changed since the first edition of *Rethinking Mathematics*, though some things stayed the same. The U.S. people elected Barack Obama president—twice—but Guantánamo is still open. Hurricanes Katrina and Sandy have come and gone, the latter maybe finally convincing people that climate catastrophe is real and upon us. The Arab Spring has sprung, the "1 percent-99 percent" divide has been lifted up by social movements, and the U.S. economy is stumbling along and suffering from its worst disaster since the 1930s—while elites continue to push policies that only exacerbate inequality.

In education, business is hot and lucrative. Schemes abound to privatize numerous aspects of schooling, but the resistance is coalescing and strengthening as well. In Chicago, the 2012 seven-day teachers' strike, led by a new union leadership committed to principles of social justice, was a dramatic victory in many ways with national implications for both education and labor. It showed that people's movements can create real solidarity between progressive organizations, communities of color, and a social justice union, and that, together, they could unite people across a city to articulate a more democratic vision of education. Across the United States, educators are collaborating and creating organizations that work with community organizations against high-stakes testing, school closings, charter proliferation, and apartheid-style education, and for critical and culturally relevant education that speaks to full human development for *all* students.

Teaching math for social justice is part of this larger effort. But we know it's not easy. High-stakes testing, accountability regimes, threats of probation/closing/reconstitution, teacher evaluations based on student test scores, and irrelevant top-down standards, all pushed by the Bush and Obama administrations and the Department of Education, make it difficult. These measures narrow the curriculum and constrain even innovative, creative teachers, let alone critical ones wanting to push back against the system. Most teachers recognize that schools are not factories, children are not widgets, and standardized test scores tell us little. Years ago, Donaldo Macedo termed this "education for stupidification," and today its manifestations have become more acute.

But there is more that has to do with math education. President Obama's 2009 Educate to Innovate initiative put $260 million into "science, technology, engineering, mathematics" (STEM) education to raise U.S. students "from the middle to the top of the pack in science and math over the next decade." The administration's Race to the Top prioritizes grants to states that push STEM education. The Gates Foundation, Microsoft, Time-Warner Cable, and many other corporate entities strongly support these plans to reclaim, once again, U.S. global economic supremacy in a period of ever-intensifying international competition for markets, lax environmental regulations, natural resources, and cheap labor supplies—all in the service of maximizing profit. In the White House's plainspoken words, the goal is to "enable them [U.S. students] to outcompete any worker, anywhere in the world." And math is a key way to do it. Or so they claim.

We think our response, as critical math teachers, is instead to position teaching and learning mathematics in the service of humanity and nature. No knowledge is neutral, and the same mathematics that can fly killer drones and cook up arcane financial derivatives that bankrupted millions can help the visually impaired navigate their world, epidemiologists stem the HIV/AIDS tide, and environmental justice activists express the dangers of fossil fuels.

Since 2007, the U.S.-based RadicalMath. org group has held five national conferences on math and social justice. Internationally, critical mathematics educators recently held the seventh

Mathematics Education and Society conference in Cape Town, South Africa, bringing together people from all continents to further the agenda of mathematics, equity, and justice. There is vital and critical work under way in the mathematics education world, and it's growing. We wish to make one small contribution by adding *Rethinking Mathematics, 2nd edition* to the mix.

As we release this edition, we want to clear up a few misconceptions. One is that the "critical math camp" suggests that social justice math be taught only to students from marginalized communities—that is, students of color and low-income/working-class students. We definitely believe students of color should learn mathematics for social justice, given the profound impact of racism on their lives, which is exacerbated by the current economic crisis and a system that increasingly regards them and their communities as disposable. But in contrast to the notion that only they should learn critical mathematics, we believe that all students, including those from schools and communities of privilege, need to study their reality and learn mathematics to understand and shape the world for the better.

Second, some think that social justice math is watered-down math. "Where is the math?" is the implicit question. That would be a powerful injustice indeed. In fact, our commitment is for students to learn the necessary math to deal with and get past the unjust gates in front them—*and* the math necessary to tear down the gates entirely. The stronger students' grasp of math, the better equipped they are to comprehend and change the world.

Third, we've heard some say that social justice math proponents believe that "relevant" math is necessary for black, brown, and low-income students, as they cannot learn math otherwise. Actually, we believe that math (and all education) should tap into who people are and build on their knowledge, culture, language, and experiences—but never stop there. Although an education focused on liberation can *start* from the realities of the young people in front of us, it should always go beyond. Global forces come into our lives in both obvious and subtle ways, and we would be foolish not to teach about them. In fact, that would be the greatest disservice we could do to our students, because we would not be preparing them to change the world, at a time when its transformation is so urgent.

## What's New in the 2nd Edition

This edition is substantially longer than the first. Although we have kept most chapters from the first edition, we have also added 15 new pieces (five that previously appeared in *Rethinking Schools*). The diversity of the additions is striking. They cover elementary, middle school, and high school; and in-school, out-of-school, adult education, after-school, and summer school settings. Writers include student teachers, experienced teachers, community organizers, and college-based teacher educators. Contributions include classroom vignettes and curriculum suggestions ("activity boxes"). The learning sites include private, alternative, and public schools, as well as community-based settings; middle- and upper-middle-class contexts to low-income ones; and all-black, all-Latina/o, all-female settings, as well as ones of mixed race and gender. Geographically, the new additions range from Canada to Brazil, San Francisco to New York, with others scattered across the country. Together, with the articles from the first edition, we hope that *Rethinking Mathematics, 2nd Edition* is a resource that teachers will use to grow as professionals and become, over time, competent and confident social justice educators. Toward that end, we dedicate it to those around the world who are committed to supporting students in reading and writing the world with mathematics.

# Introduction

I thought math was just a subject they implanted on us just because they felt like it, but now I realize that you could use math to defend your rights and realize the injustices around you. ... [N]ow I think math is truly necessary and, I have to admit it, kinda cool. It's sort of like a pass you could use to try to make the world a better place.

— **Freida, 9th grade, Chicago Public Schools**

Some students would prefer to have a dentist drill their teeth than to sit through a math class. Others view math class as a necessary but evil part of getting through school. Still others enjoy playing and working with numbers and problems.

We agree with Freida. Math is often taught in ways divorced from the real world. The alternative we propose in this book is to teach math in a way that helps students more clearly understand their lives in relation to their surroundings, and to see math as a tool to help make the world more equal and just.

In a "rethought" math class, teachers make mathematics more lively, accessible, and personally meaningful for students, who in turn learn in more depth.

The articles in this book provide examples of how to weave social justice issues throughout the mathematics curriculum and how to integrate mathematics into other curricular areas. This approach seeks to deepen students' understanding of society and to prepare them to be critical, active participants in a democracy.

The elementary school, middle school, high school, and college teachers who have contributed to this book also note the many potential benefits of such a social justice approach to mathematics. Among them:

- Students can recognize the power of mathematics as an essential analytical tool to understand and potentially change the world, rather than merely regarding math as a collection of disconnected rules to be rotely memorized and regurgitated.
- Students can deepen their understanding of important social issues, such as racism and sexism, as well as ecology and social class.
- Students can connect math with their own cultural and community histories and can appreciate the contributions that various cultures and peoples have made to mathematics.
- Students can understand their own power as active citizens in building a democratic society and become equipped to play a more active role in that society.
- Students can become more motivated to learn important mathematics.

These benefits come both when teachers reshape the mathematics curriculum with a social justice vision and when they integrate social justice mathematics across the curriculum into other subjects, such as social studies, science, health, reading, and writing.

## An Essential Tool for Understanding and Changing the World

To have more than a surface understanding of important social and political issues, mathematics is essential. Without mathematics, it is impossible to fully understand a government budget, the impact of a war, the meaning of a national debt, or the long-term effects of a proposal such as the privatization of Social Security. The same is true with other social, ecological, and cultural issues: You need mathematics to have a deep grasp of the influence of advertising on children;

the level of pollutants in the water, air, and soil; and the dangers of the chemicals in the food we eat. Math helps students understand these issues, to see them in ways that are impossible without math; for example, by visually displaying data in graphs that otherwise might be incomprehensible or seemingly meaningless.

As an example, consider racial profiling. This issue only becomes meaningful when viewed through a mathematical lens, whether or not the "viewer" appreciates that she or he is using mathematics. That is, it is difficult to declare that racial profiling occurs unless there is a sufficiently large data set and a way to examine that data. If, for example, 30 percent of drivers in a given area are African Americans, and the police stop six African American drivers and four white drivers, there is weak evidence that racial profiling exists. But if police stop 612 African American drivers and 423 whites, then there is a much stronger case.

The explanation lies in mathematics: In an area where only 30 percent of the drivers are black, it is virtually impossible for almost 60 percent of more than 1,000 people stopped randomly by the police to be black.

The underlying mathematical ideas—(dis)proportionality, probability, randomness, sample size, and the law of large numbers (that over a sufficiently large data set, the results of a probability simulation or of real-world experiences should approximate the theoretical probabilities)—all become part of the context that students must understand to really see, and in turn demonstrate, that something is amiss. Thus with a large data set, one can assert that a real problem exists and further investigate racial profiling. For youth, racial profiling may mean being "picked on," but the subtleties and implications are only comprehensible when the mathematical ideas are there. (See "Driving While Black or Brown," page 16.)

When teachers weave social justice into the math curriculum and promote social justice math "across the curriculum," students' understanding of important social matters deepens. When teachers use data on sweatshop wages to teach accounting to high school students or multi-digit multiplication to upper-elementary students, students can learn math, but they can also learn something about the lives of people in various parts of the world and the relationship between the things we consume and their living conditions. (See "Sweatshop Accounting," page 78, and "Sweatshop Math," page 256.)

Moreover, to understand some issues, students need to combine math with other subjects. For example, although the government releases unemployment figures monthly, Marilyn Frankenstein explains in "Reading the World with Math" (page 30) that how unemployment figures are reported profoundly affects one's understanding of what really goes on in our communities. Frankenstein points out that it's easy enough to figure unemployment percentages, but who gets counted as unemployed and who is—or is not—considered part of the workforce are political questions. The unemployment rate changes depending on these decisions. Thus math needs social studies, and social studies needs math.

## Connecting Math with Students' Cultural and Community Histories

Rethinking math also means using culturally relevant practices that build on the knowledge and experiences of students and their communities. Many of these approaches have been developed by teachers and then described and theorized by researchers of color, such as Gloria Ladson-Billings and William Tate. A guiding principle behind much of this work is that teachers should view students' home cultures and languages as strengths upon which to build, rather than as deficits for which to compensate. In "Race, Retrenchment, and the Reform of School Mathematics" (page 42), Tate offers the simple example of a teacher's failure to reach her students because she uses story problems that are not grounded in the students' culture; while Luis Ortiz-Franco ("Chicanos Have Math in Their Blood," page 95) encourages teachers to teach about the base 20 Mayan number system as a way to emphasize, to both Chicano students and others, that math has deep roots in indigenous cultures in the Americas. David Levine's article about the Algebra Project started by Bob Moses (see "Radical Equations," page 147) provides another example of teachers connecting with students' cultural and community histories to promote deeper student learning. Moses summarized the importance of these connections in his book on the project:

> [I]n the Algebra Project we are using a version of experiential learning; it starts with where the children are, experiences that they share. We get them to reflect on these, drawing on their common culture, then to form abstract conceptualizations out of their reflection, and then to apply the abstraction back on their experience.
>
> You can think of it as a circle or clock: At 12 noon students have an experience; at a quarter past they are thinking about it; at half past they are doing some conceptual work around their reflections; and at a quarter to they are doing applications based on their conceptual work. In the Algebra Project this movement from experience to abstraction takes the form of a five-step process that introduces students to the idea that many important concepts of elementary algebra may be accessed through ordinary experiences. Each step is designed to help students bridge the transition from real life to mathematical language and operations.
>
> Because of this connection with real life, the transition curriculum is not only experiential; it is also culturally based. The experiences must be meaningful in terms of the daily life and culture of the students. One key pedagogical problem addressed by the curriculum is that of providing an environment where students can explore these ideas and effectively move toward their standard expression in school mathematics.

## Understanding Their Power as Active Citizens

As students develop deeper understandings of social and ecological problems that we face, they also often recognize the importance of acting on their beliefs. This notion of nurturing what Henry Giroux has called "civic courage"—acting as if we live in a democracy—should be part of all educational settings, including mathematics classrooms.

*Rethinking Mathematics* spotlights several examples of student activism. These include 5th-grade Milwaukee students writing letters to social studies textbook publishers based on their mathematical analysis of slaveholding presidents and textbooks' failure to address this issue (see "Write the Truth," page 226); New York City students who measured their school space, calculated inequities, and then spoke out against these inequities in public forums (see "'With Math, It's Like You Have More Defense,'" page 129); and students who used math to convince their school administration to stop making so many obtrusive PA announcements (see the activity "Tracking PA Announcements," page 208).

## Motivated to Learn Important Math

Engaging students in mathematics within social justice contexts increases students' interest in math and also helps them learn important mathematics. Once they are engaged in a project, like finding the concentration of liquor stores in their neighborhood and comparing it to the concentration of liquor stores in a different community, they recognize the necessity and value of understanding concepts of area, density, and ratio. These topics are often approached abstractly or, at best, in relation to trivial subjects. Social justice math implicitly tells students: These skills help you understand your own lives—and the broader world—more clearly.

## Getting Started

Teachers and preservice teachers sometimes ask: How do I get started integrating social justice concepts in my math class? Our best advice is to take a little at a time. One way is to identify a concept/skill that you are teaching as part of your regular curriculum and relate it to a lesson idea in this book (or on the website for this book, www.rethinkingschools.org/math). Teach the lesson or unit and then gauge how successful it was in terms of student motivation, student understanding of the math concepts, and the deepening of the students' knowledge about the particular social issue.

Another way is to get to know your students and their communities well and listen closely to the issues they bring up. Many of our own social justice projects started from conversations with students about their lives or from knowing about issues in their communities. For example, see the activity "Environmental Hazards," page 52, in which high school students investigate contamination in their own neighborhoods.

The media are also potential sources of projects, because current issues both affect students' lives and have mathematical components that teachers can develop into social justice projects. For example, see "Home Buying While Brown or Black," page 61.

Certainly working in a school that has a conceptually strong foundational mathematics curriculum is helpful. Teachers cannot easily do social justice mathematics teaching when using a rote, procedure-oriented mathematics curriculum. Likewise a text-driven, teacher-centered approach does not foster the kind of questioning and reflection that should take place in all classrooms, including those where math is studied.

By saying this, we do not wish to imply that if teachers use a conceptually based curriculum that embraces the standards put forth by the National Council of Teachers of Mathematics (NCTM)—such as *Investigations* in the elementary grades, *Mathematics in Context* or *Connected Mathematics Project* for the middle grades, and *Interactive Mathematics Program* in high school— such a curriculum will automatically guide students towards a social justice orientation. In fact, these programs have an unfortunate scarcity of social justice connections. But a strong, conceptually based foundational curriculum can be a great asset to social justice math teaching, because it can encourage students to critique answers, question assumptions, and justify reasoning. These are all important dispositions toward knowledge that teachers can integrate into their social justice pedagogy.

Occasionally, a teacher needs to defend this kind of curriculum to supervisors, colleagues, or parents. One approach is to survey your state's math standards (or the national standards) and to find references to "critical thinking" or "problem-solving" and use those to explain your curriculum. Also, the NCTM clearly states that "mathematical connections" between curriculum and students' lives are important.

But it's important for teachers to recognize that social justice math is not something to sneak into the cracks of the curriculum. It's not something about which we should feel defensive. What we're talking about here is something that helps students learn rich mathematics, motivates them, and is really what math is all about. A social justice approach to math is the appropriate type of math for these unjust times. Other, traditional forms of math are often too abstract, promote student failure and self-doubt, and, frankly, are immoral in a world as unjust as ours. Traditional math is bad for students and bad for society.

### Views on Math and Social Justice

The two of us have been teaching math for a combined total of more than 50 years—one of us in a bilingual 5th-grade classroom in a public elementary school and the other in inner-city public middle and high schools, in alternative high schools, and at the college level. Our perspectives on teaching math for social justice have been shaped by our own involvement in movements for social justice during the past four decades—the Civil Rights Movement, anti-war movements, educational justice movements, and

other campaigns. We've also been influenced by educators such as the late Brazilian educator Paulo Freire, who argued against a "banking approach" to education in which "knowledge" is deposited into the heads of students and in favor of "problem-posing" approaches in which students and teachers together attempt to understand and eventually change their communities and the broader world.

# Math has the power to help us understand and potentially change the world.

In addition to the benefits outlined at the beginning of this introduction, an important aspect of a social justice approach to teaching math is that it must include opening up the "gates" that have historically kept students of color, women, working-class and low-income students, and students with perceived disabilities out of advanced mathematics tracks and course offerings. The Algebra Project mentioned above, for example, seeks to ensure that "gate-keeper" classes like algebra don't prevent large numbers of historically disenfranchised students from succeeding in higher education. (For information on other such projects, see Resources, page 259.)

Those who wrote for this book, and those who write for the magazine *Rethinking Schools,* are always encouraged not only to explain what they teach and why they try certain things, but to reflect on how they would do things differently next time. In that spirit we recognize that, as white male educators, our experiences have their own limitations and, if we were to do this book over, we would work harder to increase the representation of authors of color. We encourage all educators who teach math, particularly educators of color, to write about their experiences teaching math for social justice and to consider submitting articles for possible publication in *Rethinking Schools.*

## Isn't Math Teaching Neutral?

While reading these articles, some people might question whether it's appropriate to interject social or political issues into mathematics. Shouldn't math teachers and curriculum, they might say, remain "neutral?"

Simply put, teaching math in a neutral manner is not possible. No math teaching—no teaching of any kind, for that matter—is actually "neutral," although some teachers may be unaware of this. As historian Howard Zinn once wrote: "In a world where justice is maldistributed, there is no such thing as a neutral or representative recapitulation of the facts."

For example: Let's say two teachers use word problems to teach double-digit multiplication and problem-solving skills. They each present a problem to their students. The first teacher presents this one:

> A group of youth aged 14, 15, and 16 go to the store. Candy bars are on sale for 43¢ each. They buy a total of 12 candy bars. How much do they spend, not including tax?

The second teacher, meanwhile, offers a very different problem:

> Factory workers aged 14, 15, and 16 in Honduras make McKids children's clothing for Walmart. Each worker earns 43 cents an hour and works a 14-hour shift each day. How much does each worker make in one day, excluding any fees deducted by employers?

While both problems are valid examples of applying multi-digit multiplication, each has more to say as well. The first example has a subtext of consumerism and unhealthy eating habits; the second has an explicit text of global awareness and empathy. Both are political, in that each highlights important social relations.

When teachers fail to include math problems that help students confront important global issues, or when they don't bring out the underlying implications of problems like the first example here, these are political choices, whether the teachers recognize them as such or not. These choices teach students three things:

1. They suggest that politics are not relevant to everyday situations.
2. They cast mathematics as having no role in understanding social injustice and power imbalances.
3. They provide students with no experience using math to make sense of, and try to change, unjust situations.

These all contribute to disempowering students and are objectively political acts, though not necessarily conscious ones.

As high school teacher Larry Steele details in his article "Sweatshop Accounting" (page 78), the seemingly neutral high school accounting curriculum in fact approaches the world in terms of markets and profit-making opportunities. Not everything that counts gets counted, Steele says, and thus the "neutral" curriculum actually reinforces the status quo.

We believe it's time to start counting that which counts. To paraphrase Freida, the 9th grader quoted above, we need to encourage students to defend their rights and to recognize the injustices around them. By counting, analyzing, and acting, we will help students and ourselves better read the world and remake it into a more just place.

**Eric (Rico) Gutstein and Bob Peterson**

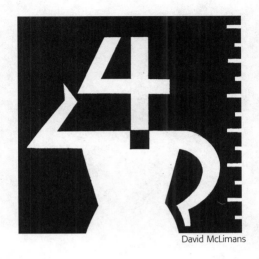

David McLimans

# Viewing Math Broadly

# CHAPTER ONE

# Teaching Math Across the Curriculum

## BY BOB PETERSON

I once read a proposal for an innovative school and it set me thinking about math. It wasn't the proposal's numbers that got my mind going, but rather the approach to structuring math into the curriculum. I disagreed with it.

The plan called for the curriculum to be divided into three areas—math/science, the arts (including fine arts and language arts), and history/philosophy. Blocks of time were set aside for a unified approach in each area.

As I mulled over the proposal and thought of my experience in a self-contained 5th-grade classroom, I realized I was uneasy with the proposed curricular divisions, specifically the assumption that science and math belong together as a unified block. It reminded me of how some elementary teachers integrate the curriculum by lumping language arts and social studies together in one strand, and math and science in another.

It also raised several questions for me. Why place math and science together and not math and social studies? What are the political and pedagogical assumptions behind such an approach? Why shouldn't reformers advocate math in all subject areas? Why not have "math across the curriculum," comparable to "writing across the curriculum"?

One reason reformers have advocated changes in how math is structured is because of the historic problems with math instruction itself: rote calculations, drill and practice ad nauseum, endless reams of worksheets, and a fetish for "the right answer." These have contributed to "number numbness" among students and, ultimately, among the general population when students become adults.

But the problem is deeper than a sterile teacher-centered and text-driven approach. "Number numbness" also has its roots in how math is segregated in schools and kept separate from the issues that confront students in their daily lives. Most students don't want to do abstract exercises with numbers or plod through text-based story problems that have them forever making change in some make-believe store. The curriculum rarely encourages students to link math and history, math and politics, math and literature—math and people. There are unfortunate consequences when math is isolated. First, the not-so-subtle message is that math is basically irrelevant except for achieving success in future math classes, becoming a scientist or mathematician, or making commercial transactions. Second, students learn that math is not connected to social reality in any substantive way. Thus students approach math in the abstract and never are encouraged to seriously consider the social and ethical consequences of how math is sometimes used in society.

Third, if students are not taught how math can be applied in their lives, they are robbed of an important tool to help them fully participate in society. An understanding of math and how numbers and statistics can be interpreted

is essential to effectively enter most debates on public issues such as welfare, unemployment, and the federal budget. For example, even though the minimum wage is higher than it's ever been, in constant dollars it is the lowest in 40 years. But you need math to understand that.

When I first began teaching more than 25 years ago, I was dissatisfied with "number numbness," but wasn't sure what to do about it. My growth as a teacher first came in the area of language arts and reading. I increasingly stressed that students should write for meaningful purposes and read books and stories that were connected to their lives. Thus I had children reading and discussing whole books, I conducted writing workshops, and I incorporated reading and writing in science and social studies.

My math, however, remained noticeably segregated from the rest of the curriculum, even though I increasingly emphasized problem-solving and the use of manipulatives.

Later, with the help of my teaching colleague Celín Arce and publications from the National Council of Teachers of Mathematics (see Resources, page 261), I began to view math as akin to language. I now believe that math, like language, is both a discipline unto itself and a tool to understand and interact with the world and other academic disciplines. Just as written and oral language help children understand their community, so can written and oral mathematics. Just as teachers stress the need for "writing across the curriculum," I believe it is important to advocate "math across the curriculum." Just as students are expected to write for meaningful purposes, they should do math for meaningful purposes.

Plans to integrate math into science are a step in the right direction. Assuming that the science curriculum is "meaningful," the teaching of mathematics will improve. But linking math with science is only a beginning and should be followed with integration of math across the curriculum. I have found that my 5th graders, for instance, are particularly interested in social issues. Thus integrating math with social studies is an effective way to bring math alive for the students.

Before I go any further, I want to make two important clarifications. First, I don't mean to imply that distinct math "mini-lessons" aren't important. They are, just as such lessons are necessary in reading and writing. I also want to make clear that integrating math with social studies does not necessarily make the teaching more student-centered or the content more concerned with issues of social justice. Those important components depend on the teacher's philosophical and pedagogical beliefs.

In recent years I have tried in a variety of ways to integrate mathematics—from the simplest understanding of number concepts to solving more complex problems—with social studies. In the interests of clarity (my classroom life is never so neatly ordered), I outline these approaches as: Connecting Math to Students' Lives, Linking Math and Issues of Equality, Using Math to Uncover Stereotypes, and Using Math to Understand History.

## Connecting Math to Students' Lives

The starting point for many teachers is to build on what students bring into the classroom, and to connect curriculum to students' lives. Math is a great way to do this. I usually start the year by having kids explore, in small groups, how math is used in their homes and communities. They scour newspapers for numbers, cut them out, glue them onto poster paper, and try to give sense to their meanings, which at times is difficult. They interview family members about how they use math and then write up their discoveries. As part of a beginning-of-the-year autobiography, they write an essay, "Numeric Me," tying in all the numbers that connect to their lives: from height and weight; to the number of brothers and sisters they have; to addresses, phone numbers, and so forth. I also ask them to write a history of their experiences in math classes, what they think about math, and why.

This process starts a yearlong conversation on what we mean by mathematics and why it's important in our lives. As the class increasingly becomes sensitive to the use of numbers and math in news articles, literature, and everyday events, our discussions help them realize that

Susan Lina Ruggles

5TH-GRADE STUDENTS SURVEY CLASSMATES TO LEARN ABOUT EACH OTHER AND THE POWER OF MATHEMATICS.

math is more than computation and definitions. It includes a range of concepts and topics—from geometry and measurements to ratios, percentages, and probability.

As part of the autobiography project we also do a timeline. We start by putting the students' birthdates and those of their parents and grandparents on a class timeline that circles the outer perimeter of my classroom (and which is used throughout the year to integrate dates that we come across in all subject areas). The students also make their own timelines—first of a typical day and then of their lives. In these activities, students use reasoning skills to figure out relations between numbers, distance, time, fractions, and decimals.

I also use another beginning-of-the-year activity that not only builds math skills but fosters community and friendship. The whole class discusses what a survey or poll is, and then brainstorms questions that they would like to ask each other. First I model how to do a survey, and as a class we graph and write about the data. Then each student surveys her or his classmates about a different topic. My students have, for example, surveyed classmates on their national origins; their favorite fast-food restaurants, music groups, or football teams; and what they think of our school's peer mediation program. Each student tabulates her or his survey data, makes a bar graph displaying the results, and reflects in writing on what she or he has learned. Later in the year students convert the data into fractions and percentages and make circle graphs. I encourage them to draw conclusions from their data and hypothesize about why the results are the way they are. They then present these conclusions orally and in writing.

This activity is particularly popular with my students, and often they want to do more extensive surveys with broader groups of people. The activity thus lays the basis for more in-depth study of polling and statistics around issues such as sampling, randomness, bias, and error. (For extensive curricular ideas on the use of polls and statistics in social studies, refer to *The Power of Numbers* curriculum published by Educators for Social Responsibility. See Resources, page 267.)

## Linking Math and Issues of Equality

To help my students understand that mathematics is a powerful and useful tool, I flood my classroom with examples of how math is used in major controversies in their community and in society at large. I also integrate math with social studies lessons to show how math can help us better understand the nature of social inequality. Kids are inherently interested in what is "fair," and using math to explore what is and isn't fair is a great way to interest them in all types of math concepts, from computation to fractions, percentages, ratios, averages, and graphing.

For example, during October and November, there is often lots of discussion of poverty and hunger in my classroom, related either to the UNICEF activities around Halloween or

issues raised by the Thanksgiving holiday. This is a good time to use simulation exercises to help children understand the disparity of wealth in the United States and around the world. In one lesson (explained in detail in "Poverty and World Wealth," page 89), I provide information on the distribution of population and wealth on the six continents and then have children represent that information using different sets of colored chips.

After I work with students so that they understand the data, we do a class simulation using a map of the world painted on our school playground. Instead of chips to represent data we use the children themselves. I tell them to divide themselves around the playground map in order to represent the world's population distribution. I then hand out chocolate chip cookies to represent the distribution of wealth. As you can imagine, some kids get far more cookies than others and lively discussions ensue. Afterwards, we discuss the simulation and write about the activity.

Not only does such a lesson connect math to human beings and social reality, it does so in a way that goes beyond paper and pencil exercises; it truly brings math to life. I could simply tell my students about the world's unequal distribution of wealth, but that wouldn't have the same emotional impact as seeing classmates in the North American and European sections of the map get so many more cookies even though they have so many fewer people.

I also use resources, such as news articles on various social issues, to help students analyze inequality. In small groups, students examine data such as unemployment or job trends, convert the data into percentages, make comparisons, draw conclusions, and make graphs. This is a great way to help students understand the power of percentages. Because they also use a computerized graph-making program, they realize how the computer can be a valuable tool.

One group, for example, looked at news stories summarizing a university report on 10,000 new jobs created in downtown Milwaukee due to commercial development. According to the report, African Americans held fewer than 8 percent of the new jobs, even though they lived in close proximity to downtown and accounted for 30 percent of the city's population. In terms of the higher-paying managerial jobs, Latinos and African Americans combined

## DISPARITIES IN WEALTH

### CARTOON TEACHING SUGGESTIONS

Ask students to describe the multiple messages contained in the cartoon below at left. Have them research the distribution of the world's wealth and which companies control major media outlets. Ask why major media corporations might not highlight disparities in wealth. Have students create graphs and cartoons to demonstrate inequality in wealth and who controls major media outlets.

held only 1 percent, while white residents, who are overwhelmingly from the suburbs, took almost 80 percent of the new managerial jobs. Using these data, students made bar and pie graphs of the racial breakdown of people in different jobs and in the city population. They compared the graphs and drew conclusions.

They then did a role play, with some students pretending to be representatives of community organizations trying to convince the mayor and major corporations to change their hiring practices. What began as a math lesson quickly turned into a heated discussion of social policy. At one point, for example, a student argued that one third of the new jobs should be given to African Americans, one third to Latinos and one third to whites, because those are the three principal ethnic groups in Milwaukee. Others,

## ACTIVITY BOX

### NO-TV WEEK MATH

During our school's annual "No-TV Week" my students survey the students in our school and use percentages and pie graphs to show, among other things, how many kids have TVs in their bedrooms, how many have cable, who can watch TV whenever they want, and who would rather spend time with their parents than watching TV.

however, disagreed. Needless to say, this led to an extensive discussion of what is "fair," why minorities held so few of the jobs created downtown, and what it would take for things to be different.

### Using Math to Uncover Stereotypes

It's important for students to be aware of whose voices they hear as they read a history book or the newspaper or watch a movie. Who gets to narrate history matters greatly, because it fundamentally shapes the readers' or viewers' perspective. We can analyze these things with kids and help them become more critical readers of books and other media. In this process math can play an important role.

I usually start with something fairly easy: I have students analyze children's books on Christopher Columbus, tabulating whose views are represented. For instance, how many times do Columbus and his men present their perspective, and how many times do the Taíno Indians present theirs? The students, using fractions and percentages, make large graphs to demonstrate their findings and draw potential conclusions. Large visual displays—bar graphs made with sticky tape, for instance—are good points of reference for discussion and analysis. Math concepts such as percentages, proportions, and comparisons can be used to help kids discuss the statistics they've uncovered and the graphs they've made.

A similar tabulation and use of percentages can be used to analyze popular TV shows for the number of "put-downs" versus "put-ups," who is quoted or pictured in newspapers, stereotypes of females in popular cartoons, who is included in textbooks, and who is represented in the biography section of the school library. (See "Libraries, Books, and Bias," page 235.)

### Using Math to Understand History

As we study history, we pay particular attention to dates and data. I try to highlight numbers that relate to social movements for equity and justice. For example, as we look at the women's struggle for equality we try to imagine what it was like for Susan B. Anthony to go to work as a teacher and get paid $2.50 a week, exactly half the salary of the male teacher she replaced. Lots can be done with such a statistic: for example, figuring out and graphing the difference on an annual or lifetime basis, or looking for wage differentials in other occupations and time periods, including the present. I have found children particularly interested in looking at wages paid to child workers, whether in coal mines or textile mills. We compare such wages to the prices of commodities at the time, to the wages of adult workers, and to the wealth that was accumulated by the owners of industry. Such historical connections can easily be linked to present-day concerns over U.S. child-labor and minimum-wage laws or to international concerns over multinational corporations exploiting child labor in Asia or Latin America to make consumer goods for worldwide markets. An in-depth description of how I have my class approach some of these global issues mathematically and socially can be found in *Rethinking Globalization: Teaching for Justice in an Unjust World* (see Resources, page 266).

One math/history connection, which can range in sophistication depending on the level of the students, is to look at who is represented in different occupations and areas of power in our society, and how that has changed over time. For example, students can figure out what percentage of the signers of the Constitution were slaveholders, common working people, women, wealthy individuals who held bonds, and so forth. A similar exercise would be to analyze U.S. presidents or the people our country has chosen to honor by putting their faces on currency and coins. Such historical number-crunching can take a contemporary turn if the students analyze, for example, the gender and racial breakdown of the U.S. House and Senate, the editors of major newspapers, or the CEOs of Fortune 500 corporations.

It's important for students to understand that such numbers are not permanent fixtures of our social structure but have changed over time as a result of social movements, such as the Civil Rights or women's movements. To demonstrate this, a teacher might have students tally the current percentage of African Americans or women in selected professional occupations and compare it to the 1960s, before the rise of affirmative action.

Another area to teach is the history of math, pointing out the contributions of various non-European cultures and civilizations to mathematical thought. Greek mathematicians, for instance, were heavily influenced by their predecessors and counterparts in Africa and Asia. Arab mathematicians inspired European Renaissance scholars.

Susan Lina Ruggles

A STUDENT TOTALS THE RESULTS OF A SURVEY HE CONDUCTED OF HIS CLASSMATES.

The Mayans were one of the first peoples to develop the concept of zero and make sophisticated mathematical calculations. I have used a unit on the base 20 Mayan counting system with my 5th graders to demonstrate such sophistication and to help students expand their understanding of place value. (See the related article "Chicanos Have Math in their Blood," page 95.)

**Conclusion**

The sophistication and complexity of the math we use in our classrooms naturally depends on the developmental level of our students. Teachers, however, too often underestimate what students are capable of doing. To the degree that I provide quality instruction, clear modeling, and purposeful activities, I am usually pleased with the enthusiasm with which my kids take on such math-based projects and with the success they have in doing them.

I have found that as a result of trying to implement "math across the curriculum"—and in particular, integrating math and social studies—my students' interest and skill in math have increased, both in terms of their understanding of basic concepts and their ability to solve problems. Furthermore, they can better clarify social issues, understand the structures of society, and offer options for better social policies.

Kids need every tool they can get to make this world—their world—a better place. Mathematics is one very important tool. ☐

# DRIVING WHILE BLACK OR BROWN

## A MATHEMATICS PROJECT ABOUT RACIAL PROFILING

BY ERIC (RICO) GUTSTEIN [FOR A MORE UPDATED VERSION USING GRAPHING CALCULATORS AND MORE RECENT DATA, SEE WWW. RETHINKINGSCHOOLS.ORG/MATH]

I did this project in 7th- and 9th-grade classes, using Chicago and Illinois data, but you can modify it for your setting. The project takes three to five class periods. Begin with a discussion of racial profiling and students' experiences with and knowledge of it. I've often started by showing the first eight minutes of the film *The Hurricane* and stopping right after Denzel Washington asks the police, "Any two will do?" as the police cars slowly surround the car he is in. The prompt to students is: "What did Denzel mean?" and "What are your experiences with driving while black or brown?"

End with a discussion of what we can and cannot know from our mathematical analysis, what are our "open" questions, what are the complexities in understanding racism, and what we can do about racial profiling.

The purpose of this project is to investigate racial profiling, also known as "Driving While Black" or "Driving While Brown." African Americans and Latinas/os have complained, filed suit, and organized against what they believe are racist police practices: being stopped, searched, harassed, and arrested because they "fit" a racial profile.

But is this true? How can we know? Can mathematics be a useful tool in helping us answer this question?

## PART 1: REVIEW BASIC PROBABILITY IDEAS

To understand racial profiling, students need to understand several concepts: randomness, experimentation, simulation, sample size, experimental and theoretical probability, and the law of large numbers (which says the more experiments you run, the closer you come to theoretical probabilities). Acquainting students with these concepts can be as simple as having pairs of students toss a coin 100 times and record their results, and then combining the data from all student pairs and examining how the combined data comes closer to a 50-50 split than do the individual pairs.

## PART 2: FIND CHICAGO'S RACIAL BREAKDOWN (DONE WITH CENSUS 2000 DATA)

Give each group of students a small bag with colored cubes to represent the racial breakdown. I used nine black cubes (for African Americans), nine tan (whites), six red (Latinas/os), and one yellow (Asians/ Pacific Islanders/Native Americans) to approximate Chicago's proportions. Do not tell students how many cubes total or of each color there are.

Have students in small groups pick a cube *without* looking, record its color, and replace the cube. Repeat this 100 times. After every 10 picks, have them record the totals on a chart (see page 18). Each line in the chart is the cumulative total of picks. Tell students that they are conducting an experiment (picking/replacing 100 times), collecting data (recording each pick), and analyzing data (determining from their simulation how many cubes of each color are in the bag, the total number of cubes, and the Chicago racial/ethnic percentages).

Make sure students record the fraction and percentage of each race/ethnicity for every 10 picks on the chart. This will help them see how the probabilities converge as the number of picks increases. If you combine all class data (e.g., on the board), you will see the law of large numbers in operation.

### QUESTIONS FOR EACH GROUP

Emphasize that you want thorough explanations for all questions.

1. What do you think is in the bag? Why? [ask this before combining all class data, then after combining]

2. What happened as you picked more, What do you think would happen if you picked 1,000 times? 1,000,000 times?

### PART 3: INVESTIGATING DWB

Here are sample Illinois data based on police reports from 1987–1997:

• In an area of about one million motorists, approximately 28,000 were Latinos.

• Over this period, state police made 14,750 discretionary traffic stops (for example, if a driver changes lanes without signaling or drives one to five miles per hour over the speed limit, police may stop her or him but do not have to).

• Of these stops, 3,100 were of Latino drivers.

Have students use what they learned in part 2 and set up their own simulation of this situation using cubes (in this example, one could use three different-colored cubes out of 100, or one out of 28, to approximate the ratio of Latino drivers). Have them pick and replace a cube 100 times, record the data, and calculate the results of simulating 100 or more "discretionary" stops.

### MORE GROUP QUESTIONS

3. What percentage of the motorists in part 3 were Latinos?

4. What percentage of the discretionary traffic stops involved Latinos?

5. How did you set up the simulation for part 3 (how many "Latino" cubes and how many total)? Why did you choose those numbers?

6. In your simulation, how many Latinos were picked out of 100 picks, and what percentage is that?

7. Do the results from your simulation experiment support the claim of racial profiling? Why or why not?

Combine individual groups' results and analyze as a whole class.

### INDIVIDUAL WRITE-UP WITH POSSIBLE QUESTIONS

1. What did you learn from this activity?

2. How did mathematics help you do this?

3. Do you think racial profiling is a problem, and if so, what do you think should be done about it?

4. What questions does this project raise in your mind?

End with whole-class discussion. ■

# DRIVING WHILE BLACK OR BROWN

## DETERMINING THE RACIAL BREAKDOWN OF THE POPULATION OF CHICAGO

| TOTAL # OF PICKS | WHITES | | | AFRICAN AMERICANS | | | LATINOS | | | ASIANS/NATIVE AMERICANS | | |
|---|---|---|---|---|---|---|---|---|---|---|---|---|
| | # | fraction | % | # | fraction | % | # | fraction | % | # | fraction | % |
| 10 | | | | | | | | | | | | |
| 20 | | | | | | | | | | | | |
| 30 | | | | | | | | | | | | |
| 40 | | | | | | | | | | | | |
| 50 | | | | | | | | | | | | |
| 60 | | | | | | | | | | | | |
| 70 | | | | | | | | | | | | |
| 80 | | | | | | | | | | | | |
| 90 | | | | | | | | | | | | |
| 100 | | | | | | | | | | | | |

# CHAPTER TWO

# Historical, Cultural, and Social Implications of Mathematics

## BY S. E. ANDERSON

The Egyptians, Chinese, and Indians used different styles of mathematical generalization in algebraic problem-solving. Pre-Hellenic algebraic proofs, such as those found in two of the most important Egyptian mathematical documents, the Ahmes Papyrus (c. 1650 B.C.) and the Moscow Papyrus (c. 1850 B.C.), while not deductively axiomatic, were and still are valid proofs.

The Egyptians created a tremendous civilization that lasted thousands of years. Egypt is in Africa, and the people who inhabit the land were and are Africans. The Egyptian civilization produced complex technological innovations and forms of communication, and engaged in an extensive interchange of goods and ideas with other people, thousands of years before they helped bring forth Greek civilization. At least half of the Greek language is African in origin and the Greek cosmological and mythological constructs were founded upon Egyptian constructs, as was Greek shipbuilding, architecture, and mathematics.

Euclid—the so-called "father" of plane geometry—spent 21 years studying and translating mathematical tracts in Egypt. Pythagoras also spent years studying philosophy and science in Egypt, and possibly journeyed East to India and/or Persia, where he "discovered" the so-called Pythmatical documents (c. 800–500 B.C.). How could a theorem whose proof was recorded in Babylonian documents dating 1,000 years before he was born be attributed to Pythagoras?

We need to shatter the myth that mathematics was or is a "white man's thing," and to show that all civilizations, though they differ and develop at different paces, have always been bound inextricably to each other. It is important that students know that Europe is not now, nor was it ever, the "civilizing center" of the world surrounded by wildness and chaos.

To further undo this Eurocentric assumption, students need to know about the constant flow of ideas and techniques into Europe from the early Greeks through the Medieval and Renaissance periods to the rise of capitalism. Certain aspects of European mathematics could not have developed had not the Europeans traded with more advanced societies. One of the most glaring examples of this is the case of the 150-year political struggle around the incorporation of the Hindu-Arabic numeral system into common usage in Europe. For a century and a half during the Medieval period (specifically around 1200–1350 A.D.), the dominant Roman Catholic Church's fear of a rising rival class, coupled with European racism and xenophobia, impeded the spread of mathematical knowledge throughout Europe. The Vatican denounced Hindu-Arabic numerals as "the work of the devil" because it viewed the widespread use of an easy way to calculate as a means by which European merchants and craftsmen would become even more independent of the Church.

The example leads into a discussion of the myth of the "Dark Ages," which asserts that because a general retrogression occurred among the European feudal elite during the Medieval period, nothing was happening intellectually anywhere else in the world. On the contrary, ideas in mathematics, science, and philosophy flourished, both inside and outside of Europe during that time. Great African and Middle-Eastern Arab scholars lived and studied in places like Toledo and Cordoba in Spain as well as in Sicily, Timbuktu, Cairo, Baghdad, and Jundishapur, which were also key centers of scholarly learning and research. The rich and complex Arab culture that dominated southern Europe, most of Africa, the Middle East, and parts of India and China during this period brought forth such intellectual centers as Caliph al-Mansur's House of Wisdom (Bait al-Hikma) in Baghdad, where documents, scholars, and researchers integrated the great astronomical studies of Indian, Chinese, Greek, and Babylonian scholars.

Bait al-Hikma was also a research university out of which a key mathematician evolved: Mohammed ibn-Musa al-Khwarizmi (c. A.D. 825–?). He authored two foundational mathematics texts: The first, *Hisab al-djabr wa-al Muqabala* or *The Science of Reduction and Cancellation* not only gave Europe its first systematic approach to algebra, but also was the source of the name for the subject matter (al-djabr or "algebra"). His second book (now found only in the Latin original), *Algorithmi de Numero Indorum*, explains the Indian origins of the numeral system. Subsequent European translations of this book attributed the system to him; hence, schemes using these numerals came to be known as "algorithms" (a corruption of the name "al-Khwarizmi"), and the numerals became known as "Arabic numerals." ❑

This article is adapted from a piece that first appeared in the *Journal of Negro Education* and later in "Worldmath Curriculum: Fighting Eurocentrism in Mathematics" in *Ethnomathematics: Challenging Eurocentrism in Mathematics Education* edited by Arthur B. Powell and Marilyn Frankenstein.

# CHAPTER THREE

# 'I Thought This U.S. Place Was Supposed to Be About Freedom'

## Young Latinas Engage in Mathematics and Social Change to Save Their School

BY MAURA VARLEY GUTIÉRREZ

I sat and talked with a group of 5th-grade Latina girls, their voices filled with urgency and excitement. We were gathered for our twice-weekly meeting of the after-school girls' mathematics club at Agave Elementary School* in the U.S. Southwest. I was the facilitator of the group, conducting participatory research as a part of my graduate studies. The seven Latinas represented a range of mathematical abilities and confidence, as well as language skills (some bilingual in Spanish and English, some English dominant). On this January day, our usual opening activity of sharing stories from our lives turned to the subject of the school's potential closure, which had been proposed by the school district the week before.

At a community meeting attended by several of us the evening before, the district officials laid out their argument for closing four schools, claiming that they used an objective, neutral, mathematical procedure for determining which schools to close. The girls' discussion covered a variety of concerns about the potential impact of the closing on their community. "It's not fair, though!" Zara insisted, drawing attention to what the school district left out of its arguments: the social impact of the school closings. This led to a heated discussion that included sharing personal stories and knowledge about the community, voicing concern for others, expressing disagreement with the district's arguments, and suggesting ideas for how to save the school.

After a semester of building community within the group and investigating topics such as prison vs. education spending in our state, we had planned to explore school safety. However, given the news of the school closures, we shifted the club's focus because, as Margarita reasoned, "What point is there to do school safety if there's gonna be no school?"

Between January and the final school board vote in April, the Agave community mobilized to defend their neighborhood school—a school that had been attended by generations of primarily Latino families. Zara's assertion that the closure was "not fair" seemed an ideal place to begin to push our conversation to how we might specifically use mathematics as a tool for participating in this mobilization. Our math-based participation in the community movement proved to be a powerful experience for both the girls and for me as an educator.

## The Mathematical Investigations

The district argued that Agave was a good candidate to be closed because of its proximity to the proposed receiving school (1.3 miles) and its "underperforming" label based on standardized tests. When the girls questioned these arguments, I helped them shape their ideas into mathematical problems that they could solve. This evolved into a unique contribution to the community movement to save the school. We focused on alternative analyses of test score data and the school's achievement label, and on the walk students would have to take to the new school.

During our discussion, the girls asked, "How can they say we are un[der]achieving?" I helped the girls break this broad question into smaller parts that could be solved by students with a range of mathematical experience. I provided the students with details about the district's argument and introduced the girls to a district website with several years of test scores.

Zara decided to analyze test score data over time. She saw a trend toward improvement despite the school's failure to meet NCLB goals. She also disaggregated the scores by gender in order to show a trend of higher test scores for girls. Then she and Alma worked together to represent these trends in a double bar graph. In a letter to a board member, Zara wrote, "As you can see, every year we keep going up or are almost the same. For example, 40.4 percent of the girls passed the AIMS test three years ago, and two years ago 54.3 percent of the girls passed the test. Three years ago 26.5 percent of boys passed the test, and two years ago 40.4 percent of boys passed the test. Please think about these things when you vote on Tuesday." Margarita asked the board to put the test scores into a larger context. Referring to the strong partnership between Agave and the local community center, she noted that some schools "are really high [achieving on tests] and all, but do they have things that prevent kids from going to bad places, from doing bad things? Here [at Agave] we do."

After finding alternative ways to evaluate Agave's achievements, we examined the impact on the community of students attending a different school. The fact that all of Agave's students would be transferred to a single school (Samota) was described by the district superintendent as "seamless." The voices of community members and conversations in the girls' group revealed that the move would be anything but seamless for the families of Agave. While busing would be provided for elementary-aged students who lived more than 1.5 miles from their school, most of the students who lived close to Agave would be left without school-sponsored transportation.

The girls and I decided to investigate the walk to Samota in several ways. We took the walk between the two schools ourselves, timing our journey and documenting it with videos and photos. We walked 13 blocks in 40 minutes. Then we created a map to determine our rate for walking. Most of the girls began to test whether two minutes per block was a reasonable estimate. Some solved this by multiplying two times 13. Others used repeated addition by drawing a series of 13 connected rectangles to represent the blocks; they put a two in each rectangle to represent the minutes. When a rate of two minutes per block resulted in a total walking time of only 26 minutes, the girls tried three minutes per block. They found that rate resulted in a total walking time of 39 minutes, only one minute less than our actual time. Zara decided that the extra minute was accounted for by the time used to take pictures. Drawing on her experience, Zara argued that her solution was sensible.

Having been exposed to survey work related to the 2008 presidential primary elections, Margarita suggested we ask students "what they think about the school closing." She then predicted, "I bet you it's gonna be a high number," and suggested that "if you just ask a whole bunch of people, then we can say how much people agreed that it shouldn't close."

Alma suggested a second survey question: "How about we [ask about] the number

of kids who walk to school, take the bus, and have their parents drive them to school?" The survey revealed that approximately half of the students who attended Agave walked to school, and walking was often their only available form of transportation. For many families, walking

Michael Duffy

to Agave worked because it was a neighborhood school, while walking to the new downtown school would involve additional dangers and a significantly longer walk. Michelle described the walk by saying, "When we went to go take the walk it seemed pretty dangerous because of all of the construction sites. And some kids might go through there, and it might not be a pretty picture."

We decided to also include a question that collected data on the location of students' homes in order to calculate the average walking time it would take students to walk to the new school, using the rate we had determined from our walk. Michelle created a list of the intersections where students reported their homes were located. Then Margarita and Vane located these intersections on our wall map of the neighborhood, and all of the girls determined distances and calculated potential walking times to Samota based on the estimated walking rate. Following my suggestion, they used a line plot to represent the spread of all of the walking times.

## The problems encouraged collaboration because of the many skills necessary.

In general, the structure of the mathematical problems involved multiple entry points and could be solved in a variety of ways, which invited participation from girls with varying levels of understanding of mathematical concepts. The mathematical skills included quantifying, determining rate, calculating percent, analyzing data, and determining an appropriate representation. The problems encouraged collaboration because of the many skills necessary to complete our complex tasks, including designing surveys and collecting, organizing, and representing large amounts of data. Vanessa described the importance of this collective mathematical activity: "When you work together and you find out what [a peer's] strategy is, you might be interested in that one and you might find a new way to do it." Having the whole group contribute to the counterarguments allowed for a collective sense of arguing against the

closures. This supported the participation of all of the girls, rather than only those who tended to be successful in the traditional classroom.

### 'People Together Make Such a Big Difference'

The girls participated in and contributed to a variety of aspects of the movement, including writing letters to board members, speaking at community forums, and attending board meetings. We shared the results of our mathematical investigations in a digital story that we produced and presented at a forum attended by 600 community members, including students, teachers, parents, business owners, and city and state officials. Although the district conducted this particular forum, the community organized a rally prior to the meeting that involved singing, speeches, and forming a human chain around the entire school. Having participated in events as a part of the larger community movement, the girls learned about the arguments of both sides. They saw community members participate, they spoke in front of the board members, and they saw how decisions were made. In effect, the girls learned about the system they sought to change.

It was apparent from the community forces that organized to defend this primarily working-class, Latino community school that a vote to close the school could have negative political consequences for the elected board members. In April, despite intense pressure from the district superintendent, the board members ultimately voted *not* to close the schools. Margarita described what she learned from the project: "I learned that if you work together you can do something—you can make a difference. ... I would say teamwork is important because people together make such a big difference, not just one person."

Throughout their reflections and conversations, the girls expressed a sense of responsibility to give voice to younger students who might not otherwise have a say in the process. Once again, Margarita explained:

> As 5th graders they say we're role models and [as] role models, we should be able to help and speak up. It's like you're being a role model for so many people. And you're showing them that if you speak up you can make a difference. ... And then if [the possibility of school closure] ever comes across when we're not in this school anymore and it happens again, they know what to do.

## 'A Chance to Speak Your Mind'

Because this was one experience in these girls' lives, I can make no claims of lasting influence; because it did not happen within their mathematics classroom, I cannot verify shifts in classroom learning. However, I am convinced that, as educators, we must continually find ways to give students experiences of this sort within classrooms and other educational settings. If the primary goal of education is to engage students to become active citizens prepared to participate in social change, then connecting classroom learning with community movements becomes essential.

Despite the implementation challenges in our increasingly restrictive educational environment, Margarita's voice offers a window into the possibilities of critical education. She came to me one day in the middle of this journey and said that she needed to get her ideas out, writing the following: "I finally am in a group that gives you a chance to speak your mind when something very important is happening (which everyone should feel like they spoke their mind, but not everyone has experienced that yet. I finally have after nine years). I thought this U.S. place was supposed to be about freedom and people speaking what they felt." ☐

This work was part of the research agenda of CEMELA (Center for the Mathematics Education of Latinas/os), supported by the National Science Foundation, grant ESI-0424983. The views expressed here are those of the author and do not necessarily reflect the views of NSF.

*All names are changed to protect anonymity of research participants and research site.

CHAPTER FOUR

# Rethink Mathematics and Its Intersection with Race

## INTERVIEW WITH DANNY MARTIN, BY ERIC (RICO) GUTSTEIN

Danny Martin is a professor of education and mathematics at the University of Illinois at Chicago. He studies the mathematical and racial identity development among African American learners and is the author of *Mathematics Success and Failure Among African American Youth* (2000) and editor of *Mathematics Teaching, Learning, and Liberation in the Lives of Black Children* (2009). In the latter book and throughout his work, Martin frames race as a sociopolitical and historical construction, and examines everyday, institutional, and structural racism in relation to how African American students develop identities as mathematical learners. For him, the goals of mathematics education, research, and policy are the "empowerment and liberation from oppression for marginalized learners." As his colleague and friend, having worked together in the same department for eight years, I interviewed him so that readers of *Rethinking Mathematics* would have the opportunity to hear his voice and engage in "dialogue" with him.

**Gutstein**

What are your main concerns with respect to math education?

**Martin**

A big concern for me is helping to create meaningful mathematical experiences for children from different communities. I'm concerned about all children, but I'm particularly concerned that black children experience mathematics in relevant and meaningful ways in relationship to their other life circumstances.

That meaning is not just determined by me, but also by the relevance of math in their circumstances and situations. We need to build connections from the math that they do in school to the practices of parents, caregivers, or grandmothers.

My concern is making sure that black children are better respected, valued, humanized in how we talk about them, in the way that we report outcomes, talk about their experiences. We need to understand black children's experiences as black children, not necessarily in relationship to white or Asian American children (using those labels always in a very simplified way).

**Gutstein**

You've written, with Ebony McGee, that, "We argue that a critical and liberatory mathematics education is one tool that African Americans can use to understand and challenge the world around them." Can you say more?

**Martin**

Ebony and I were pulling on a longer legacy of scholars, activists, and everyday people who have been linking these things all along—mathematics learning and participation to ideas of liberation and emancipation—as well as drawing on contemporary and historical writers, like W. E. B. Du Bois, to some degree Booker T. Washington, Carter G. Woodson, and others. That's not new. I've been inspired by work on critical agency, and having students use math for social justice-oriented reasons, where "social justice" is reflected in their own beliefs about what's just and unjust and what's important to them.

DANNY MARTIN

We were just doing our small part to change the discourse and link mathematics to things other than jobs, degrees, certificates, hoops, and hurdles. Those things are important, but clearly there are bigger and broader things.

## Gutstein

You've said from your reading of slave narratives and other records that black people have historically used mathematics for social justice purposes. Can you share what you've learned about this?

## Martin

I had not considered going back to slave and ex-slave narratives to seriously analyze our history and try to understand what the struggles were like and the words and narratives of people who had been in and come out of very tough, dehumanizing conditions. And then coming out of that to still voice the need for math or ciphering was pretty amazing. I couldn't ignore that and had to link that to my overall position and stance in terms of my arguments about black children and mathematics.

There are social justice arguments about math and its relationship to liberation and struggle and oppression—but this is not new. This goes back decades and hundreds of years, and it's always been there because of our struggles.

In looking at those ex-slave narratives, it's profound, humbling, and informative to read about the value and importance of mathematics to people's humanity coming out of slavery, their quest for citizenship, for full rights, and their expressing desires to read and write, to do math. That's a very powerful thing.

For example, in Heather Williams' book, *Self-Taught: African American Education in Slavery and Freedom*, she recounts how ex-slaves had to learn math so that they wouldn't be cheated as they entered the paid labor force. She tells a story of a Georgia plantation owner who drove away teachers, and the freed people concluded that he did so because he didn't want them

to know enough math to keep track of their earnings.

Then it struck me, in looking at those narratives—many of the middle school students who were a part of my research years ago and who often cited "not being cheated" as a reason to learn math, were saying the thing that those ex-slaves were saying about liberation hundreds of years ago.

## Gutstein

How do you think what you've just laid out affects how readers should view teaching math for social justice and the present?

## Martin

To have that sense of history prevents a fantasizing of social justice, it's like the latest and greatest. No, it's not the latest and the greatest. There's a longer legacy. We're drawing on these legacies that we should acknowledge and try to understand and build on, be respectful of, and learn from.

If those ex-slaves could clamor for numeracy in those kinds of conditions, there is no way that we can't get contemporary kids to clamor for numeracy and use it effectively in the conditions of today that are not disconnected from the conditions of 50, 75, 125, 150, 200 years ago. History is a good teacher.

## Gutstein

What do you think are the main challenges facing teachers who want to teach mathematics for social justice?

## Martin

A necessary step is critical self-reflection. That's often difficult given one's identities, locations, and positions in relationship to the people that you're working with. In the case of white teachers, how do they understand and confront whiteness as an identity and ideology so it doesn't promote guilt or paralysis?

If your underlying beliefs and ideologies are compatible with the needs of black children and you're a white teacher, I think that's good. If you're willing to engage your own motives, beliefs, and values, that gets around the "who can do it?" question in terms of race.

You need an appreciation for justice/injustice so that teaching math for social justice is not formulaic. It's more than what you do as a performance, it's deeper than that, and it can't be scripted.

Your social justice may not be what the kids consider to be social justice. You might identify issues and oppressions that your students accept for different reasons. You might be trying to fight a system that students think is the gateway to some destination in life without realizing how that same system could be oppressing. What do you do when you have to help students shed light on that without being oppressive against their own beliefs?

If you're really committed to social justice for black students then you've got to understand what their needs are based on how they see themselves and are being positioned in the world and positioned in relationship to you. So I think there are lots of challenges on the personal level.

### Gutstein

What do you want to say to those who teach mathematics to black students, children, and youth?

### Martin

I want people to engage in some deeper reflection on their basic assumptions about black children and know that those assumptions are then going to lead to what you do. If your assumptions are problematic then probably what you're doing is problematic, even though it might be well intentioned, so the starting place is important.

People need to realize that black children are as complex and human as all children, and we don't want to reduce their identities, for black children are also children of the world. Even in places that we consider to be oppressive to black children, their reach can extend pretty far. They can see similarities and differences in their life circumstances with children across the globe. They can go on a laptop computer and see somebody in South Africa or somewhere else doing something that may or may not look like what they do.

### Gutstein

How do you see the issues you've been concerned about, specifically racism and the racialized experiences of students learning math, affecting teaching and learning math for social justice?

### Martin

Race tends to be that very thorny, unsettling place where many folks don't want to go. To the degree that we don't go there and leave it under-conceptualized, we don't talk about it in deep and meaningful ways in relationship to math learning, We do ourselves a disservice because it lingers as though it explains it. We have to reframe that so that when we talk about race, class, and gender intersections, that race isn't muted, or if it is on the table, it doesn't prevent things from moving forward.

I'd like people to not just rethink mathematics, but to rethink mathematics and its intersection with race as one kind of rethinking. ☐

# CHAPTER FIVE

# Reading the World with Math

## Goals for a Criticalmathematical Literacy Curriculum

### BY MARILYN FRANKENSTEIN

When my students examine data and questions, such as the ones on unemployment in the box on page 34, they are introduced to the four goals of criticalmathematical literacy:

1. Understanding the mathematics.
2. Understanding the mathematics of political knowledge.
3. Understanding the politics of mathematical knowledge.
4. Understanding the politics of knowledge.

Clearly, calculating the various percentages for the unemployment rate requires goal number 1, an understanding of mathematics. Criticalmathematical literacy goes beyond this, to also include the other three goals mentioned above.

The mathematics of political knowledge is illustrated here by reflecting on how the unemployment data deepen our understanding of the situation of working people in the United States. The politics of mathematical knowledge involves the choice of who counts as unemployed. In class, I emphasize that once we decide which categories make up the numerator (number of unemployed) and the denominator (total labor force), changing that fraction to a decimal fraction and then to a percent does not involve political struggle—it involves understanding the mathematics. But making the decision of who counts where does involve political struggle—so the unemployment rate is not a neutral description of the situation of working people in the United States. And this discussion generalizes to a consideration of the politics of all knowledge.

In this article, I develop the meaning of these goals, focusing on their interconnected complexity. Underlying these ideas is my belief that the development of self-confidence is a prerequisite for all learning, and that self-confidence develops from grappling with complex material and from understanding the politics of knowledge.

## Goal 1: Understanding the Mathematics

I teach at the College of Public and Community Service at the University of Massachusetts-Boston. My students are primarily working-class adults who did not receive adequate mathematics instruction when they were in high school. Almost all my students know how to do basic addition, subtraction, multiplication, and division, although many would have trouble multiplying decimal fractions, adding fractions or doing long division. All can pronounce the words, but many have trouble succinctly expressing the main idea of a reading. Almost all have trouble with basic math word problems. Most have internalized negative self-images about their knowledge and ability in mathematics. In my beginning lessons I have students read excerpts where the main idea is supported by numerical

# THE WAR IN IRAQ

### CARTOON TEACHING SUGGESTIONS

Have students calculate the cost of health insurance per family based on the data in this cartoon. Have them research the costs of other consumer items or services and create cartoons to compare those costs to that of the war in Iraq. (See pages 40 and 114 for related activities.)

details and where the politics of mathematical knowledge is brought to the fore. Then the curriculum moves on to the development of the Hindu-Arabic place-value numeral system, the meaning of numbers, and the meaning of the operations.

I start lessons with a graph, chart, or short reading which requires knowledge of the math "skill of the day." When the discussion runs into a question about a math skill, I stop and teach that skill. This is a nonlinear way of learning basic numeracy, because questions often arise that involve future math topics. I handle this by previewing: The scheduled topic is formally taught, while other topics are also discussed. In this way students' immediate questions about other topics are answered, and when the formal time comes for those topics in the syllabus, students will already have some familiarity with them. For example, if we are studying the meaning of fractions and find that in 1985, two hundredths of the U.S. Senate was female, we usually preview how to change this fraction to a percent. We also discuss how no learning is linear, and how all of us are continually reviewing, re-creating, and previewing in the ongoing process of making meaning.

Further, there are other aspects about learning which greatly strengthen students' understandings of mathematics:

- Breaking down the dichotomy between learning and teaching mathematics.
- Considering the interactions of culture and the development of mathematical knowledge.
- Studying even the simplest of mathematical topics through deep and complicated questions.

These are explained in more detail below.

**Breaking Down the Dichotomy Between Learning and Teaching Mathematics.** When students teach, rather than explain, they learn more mathematics, and they also learn about teaching. They are then empowered to learn even more mathematics. As humanistic, politically concerned educators we often talk about what we learn from our students when we teach. Educator Peggy McIntosh goes so far as to define teaching as "the development of self through the development of others."

Certainly when we teach, we learn about learning. I also introduce research on math education, so that students can analyze for themselves why they did not previously learn mathematics. I argue that learning develops through teaching and through reflecting on teaching and learning. So students' mathematical understandings are deepened when they learn about mathematics teaching as they learn mathematics. Underlying this argument is Paulo Freire's concept that learning and teaching are part of the same process, and are different moments in the cycle of gaining existing knowledge, re-creating that knowledge, and producing new knowledge.

Students gain greater control over mathematics problem-solving when, in addition to evaluating their own work, they can create their own problems. When students can understand which questions it makes sense to ask from given numerical information, and can identify decisions that are involved in creating different kinds of problems, they can more easily solve problems others create. Further, criticalmathematical literacy involves both interpreting and critically analyzing other people's use of numbers in arguments. To do the latter you need practice in determining what kinds of questions can be asked and answered from the available numerical data, and what kinds of situations can be clarified through numerical data.

Freire's concept of problem-posing education emphasizes that problems with neat, pared-down data and clearcut solutions give a false picture of how mathematics can help us "read the world." Real life is messy, with many problems intersecting and interacting. Real life poses problems with solutions that require dialogue and collective action. Traditional problem-solving curricula isolate and simplify particular aspects of reality in order to give students practice in techniques. Freirian problem-posing, on the other hand, is intended to reveal the interconnections and complexities of real-life situations, even if the specific problems are not solved.

A classroom application of this idea is to have students create their own reviews and tests. In this way they learn to grapple with mathematics pedagogy issues such as: What are the key concepts and topics to include in a review of a particular curriculum unit? What are clear, fair, and challenging questions to ask in order to evaluate understanding of those concepts and topics?

**Considering the Interactions of Culture and the Development of Mathematical Knowledge.** When we are learning the algorithm for comparing the size of numbers, I ask students to think about how culture interacts with mathematical knowledge in the following situation: Steve Lerman was working with two 5-year-olds in a London classroom. As he recounted:

> [They] were happy to compare two objects put in front of them and tell me why they had chosen the one they had [as bigger]. However, when I allocated the multilinks to them (the girl had eight and the boy had five) to make a tower ... and I asked them who had the taller one, the girl answered correctly but the boy insisted that he did. Up to this point the boy had been putting the objects together and comparing them. He would not do so on this occasion, and when I asked him how we could find out whose tower was the taller he became very angry. I asked him why he thought that his tower was taller and he just replied "Because IT IS!" He would go no further than this and seemed to be almost on the verge of tears.

At first students try to explain the boy's answer by hypothesizing that each of the girl's links was smaller than each of the

boy's, or that she built a wider, shorter tower. But after reading the information, they see that this could not be the case, since the girl's answer was correct. We speculate about how the culture of

## Real life poses problems with solutions that require dialogue and collective action.

sexism—that boys always do better or have more than girls—blocked the knowledge of comparing sizes that the boy clearly understood in a different situation.

**Studying Mathematical Topics Through Deep and Complicated Questions.** Most educational materials and learning environments in the United States, especially those labeled as "developmental" or "remedial," consist of very superficial, easy work. They involve rote or formulaic problem-solving experiences. Students get trained to think about successful learning as getting high marks on school or standardized tests. I believe this is a major reason that what is learned is not retained and not used. Further, making the curriculum more complicated, so that each problem contains a variety of learning experiences, teaches in the nonlinear, holistic way in which knowledge is developed in context. This way of teaching leads to a clearer understanding of the subject matter.

**Example:** In the text below, writers Holly Sklar and Charles Sleicher demonstrate how numbers presented out of context can be very misleading. I ask students to read the text, which is taken from a February 1990 letter published in *Z* magazine, and discuss the calculations Sklar and Sleicher performed to get their estimate of the U.S. expenditure on the 1990 Nicaraguan election ($17.5 million divided by the population of Nicaragua equals $5 per person). This reviews their understanding of the meaning of the operations. Then I ask the students to consider the

# ACTIVITY BOX

## USING MATH TO TAKE A CRITICAL LOOK AT HOW THE UNEMPLOYMENT RATE IS DETERMINED

BY MARILYN FRANKENSTEIN

In the United States, the unemployment rate is defined as the number of people unemployed divided by the number of people in the labor force. Here are some figures from December 1994. (All numbers in thousands—that is, these figures are one thousandth of the actual total—rounded off to the nearest hundred-thousand.)

1.  101,400 employed full time
2.  19,000 employed part time, want part-time work
3.  4,000 employed part time, want full-time work
4.  5,600 not employed, looked for work in last month, not on temporary layoff
5.  1,100 not employed, on temporary layoff
6.  400 not employed, want a job now, looked for work in last year, stopped looking because discouraged about prospects of finding work
7.  1,400 not employed, want a job now, looked for work in last year, stopped looking for other reasons
8.  60,700 not employed, don't want a job now (adults)

In your opinion, which of these groups should be considered unemployed? Why?

Which should be considered part of the labor force? Why?

### DO THE MATH
Given your selections, calculate the unemployment rate in 1994.

### FOR DISCUSSION
The U.S. official definition counts 4 and 5 as unemployed and 1 through 5 as part of the labor force, giving an unemployment rate of 5.1 percent. If we count 4 through 8 plus half of 3 as unemployed, the rate would be 9.3 percent. Further, in 1994 the Bureau of Labor Statistics stopped issuing its U-7 rate, a measure which included categories 2, 3, 6, 7, and 8, so now researchers will not be able to determine "alternative" unemployment rates.

complexities of understanding the $17.5 million expenditure. This deepens their understanding of how different numerical descriptions illuminate or obscure the context of U.S. policy in Nicaragua, and how in real life just comparing the size of the numbers, out of context, obscures understanding.

On the basis of relative population, Holly Sklar has calculated that the $17.5 million U.S. expenditure on the Nicaraguan election is $5 per person and is equivalent to an expenditure of $1.2 billion in the United States. That's one comparison all right, but it may be more relevant to base the comparison on the effect of the expenditure on the economy or on the election, i.e., to account for the difference in per capita income [between U.S. and Nicaraguan citizens], which is at least 30/1 or an equivalent election expenditure in the United States of a staggering $36 billion! Is there any doubt that such an expenditure would decisively affect a U.S. election?

### Goal 2: Understanding the Mathematics of Political Knowledge

I argue, as do educators Paulo Freire and Donaldo Macedo, that the underlying context for critical adult education, and criticalmathematical literacy, is "to read the world." To accomplish this goal, students need to learn how mathematics skills and concepts can be used to understand the institutional structures of our society. This happens through:

- Understanding the different kinds of numerical descriptions of the world (such as fractions, percents, graphs) and the meaning of the sizes of numbers.
- Using calculations to follow and verify the logic of

someone's argument, to restate information, and to understand how raw data are collected and transformed into numerical descriptions of the world. The purpose underlying all calculations is to better understand information and arguments, and to be able to question the decisions that were involved in choosing the numbers and the operations.

**Example:** I ask students to create and solve some mathematics problems using the information in the following article, which appeared in the April 29–May 5, 1992, edition of *In These Times* magazine. Doing the division problems implicit in this article deepens understanding of the economic data and shows how powerfully numerical data reveal the structure of our institutions.

It may be lonely at the top, but it can't be boring—at least not with all that money. Last week the federal government released figures showing that the richest 1 percent of American households was worth more than the bottom 90 percent combined. And while these numbers were widely reported, we found them so shocking that we thought they were worth repeating. So here goes: In 1989 the top 1 percent of Americans (about 934,000 households)

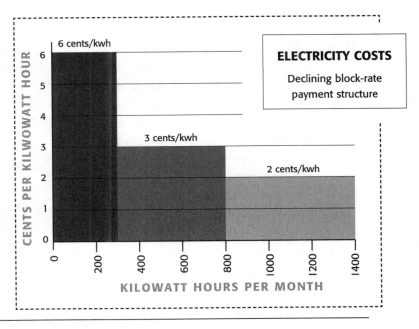

ELECTRICITY COSTS

Declining block-rate payment structure

combined for a net worth of $5.7 trillion; the bottom 90 percent (about 84 million households) could only scrape together $4.8 trillion in net worth.

**Example:** Students practice reading a complicated graph and solving multiplication and division problems in order to understand how particular payment structures transfer money from the poor to the rich.

The "Rate Watcher's Guide," produced by the Environmental Action Foundation, details why under declining block-rate structures low-income citizens who use electricity only for basic necessities pay the highest rates, and large users with luxuries such as trash compactors, heated swimming pools, or central air-conditioning pay the lowest rates. A 1972 study conducted in Michigan, for example, found that residents of a poor urban area in Detroit paid 66 percent more per unit of electricity than did wealthy residents of nearby Bloomfield Hills. Researchers concluded that "approximately $10 million every year leave the city of Detroit to support the quantity discounts of suburban residents." To understand why this happens, use the graph on page 35 to:

1. Compute the bill of a family which uses 700 kilowatt-hours (kwh) of electricity per month and the bill of a family which uses 1400 kwh.
2. Calculate each family's average cost per kwh.
3. Discuss numerically how the declining block-rate structure functions and what other kinds of payment structures could be instituted. Which would you support and why?

**Example:** Students are asked to discuss how numbers support Helen Keller's main point and to reflect on why she sometimes uses

fractions and other times uses whole numbers. Information about the politics of knowledge is included as a context in which to set her views. Historian Howard Zinn writes:

Although Helen Keller was blind and deaf, she fought with her spirit and her pen. When she became an active socialist, a newspaper wrote that "her mistakes spring out of the ... limits of her development." This newspaper had treated her as a hero before she was openly socialist. In 1911, Helen Keller wrote to a suffragist in England: "You ask for votes for women. What good can votes do when ten elevenths of the land of Great Britain belongs to 200,000 people and only one eleventh of the land belongs to the other 40 million people? Have your men with their millions of votes freed themselves from this injustice?"

**Example:** Students are asked to discuss what numerical understandings they need in order to decipher the following data. They see that a recognition of how very small these decimal fractions are, so small that watches cannot even measure the units of time, illuminates the viciousness of time-motion studies in capitalist management strategies.

Economist Harvey Braverman gives samples from time-and-motion studies conducted by General Electric, which were published in a 1960 handbook to provide office managers with standards by which clerical labor should be organized:

| Opening and Closing Desk Drawers | Minutes |
|---|---|
| Open side drawer of standard desk | 0.014 |
| Open center drawer | 0.026 |
| Close side drawer | 0.015 |
| Close center drawer | 0.027 |

| Chair Activity | Minutes |
|---|---|
| Get up from chair | 0.039 |
| Sit down in chair | 0.033 |
| Turn in swivel chair | 0.009 |

## Goal 3: Understanding the Politics of Mathematical Knowledge

Perhaps the most dramatic example of the politics involved in seemingly neutral mathematical descriptions is the choice of a map to visualize that world. Any two-dimensional map of our three-dimensional Earth will contain mathematical distortions. (See the Mercator and Peters maps, pages 192 and 193.)

The political struggle/choice centers around which of these distortions are acceptable to us, and what other understandings of ours are distorted by these false pictures. For example, the world map with which most people are familiar, the Mercator map, greatly enlarges the size of "Europe" and shrinks the size of Africa. Most people do not realize that the area of what is commonly referred to as "Europe" is smaller than 20 percent of the area of Africa. Created in 1569, the Mercator map highly distorts land areas, but preserves compass direction, making it very helpful to navigators who sailed from Europe in the 16th century.

When used in textbooks and other media, combined with the general (mis)perception that size relates to various measures of so-called "significance," the Mercator map distorts popular perceptions of the relative importance of various areas of the world. For example, when a U.S. university professor asked his students to rank certain countries by size they "rated the Soviet Union larger than the continent of Africa, though in fact it is much smaller," because they were associating "power" with size.

Political struggles to change to the Peters projection, a more accurate map in terms of land area, have been successful with the U.N. Development Program, the World Council of Churches, and some educational institutions. However, anecdotal evidence from many talks I've given around the world suggests that the Mercator is still widely perceived as the way the world really looks.

> "In order for us as poor and oppressed people to become a part of a society that is meaningful, the system under which we now exist has to be radically changed. This means that we are going to have to think in radical terms.
>
> I use the term "radical" in its original meaning—getting down to and understanding the root cause. It means facing a system that does not lend itself to your needs and devising means by which you change that system.
>
> That is easier said than done. But one of the things that has to be faced is, in the process of wanting to change that system, how much have we got to do to find out who we are, where we have come from and where we are going. ... I am saying, as you must say too, that in order to see where we are going, we not only must remember where we have been, but we must understand where we have been."
>
> **— Ella Baker,** 1969

As Denis Wood writes in his book *The Power of Maps,* "The map is not an innocent witness ... silently recording what would otherwise take place without it, but a committed participant, as often as not driving the very acts of identifying and naming, bounding, and inventorying it pretends to no more than observe."

In a variety of situations, statistical descriptions don't simply or neutrally record what's out there. There are political struggles/choices involved in: which data are collected, which numbers represent the most accurate data, which definitions should guide how the data are counted, which methods should guide how the data are collected, which ways the data should be disaggregated, and which are the most truthful ways to describe the data to the public.

**Example:** To justify the Eurocentric argument that the Native American population could not have been so great, various "scholars" have concluded that about one million people were living in North America in 1500. Yet, as related in *The State of Native America,* edited by Annette Jaimes, other academics have argued, on the basis of burial mound archeology and other evidence, that the population of the Ohio River Valley

## Statistical descriptions don't simply or neutrally record what's out there.

alone had been that great, and that "a pre-contact North American Indian population of 15 million is perhaps the best and most accurate working number available." Admitting the latter figure would also require admitting extensive agricultural institutions, as opposed to the less reliable hunting and gathering. Cultivators of land are "primarily sedentary rather than nomadic … and residents of permanent towns rather than wandering occupants of a barren wilderness."

**Example:** In 1988, the U.S. Census Bureau introduced an "alternative poverty line," changing the figure for a family of three from $9,453 to $8,580, thereby preventing 3.6 million people whose family income fell between those figures from receiving food stamps, free school meals, and other welfare benefits. At the same time, the Joint Economic Committee of Congress argued that "updating the assessments of household consumption needs … would almost double the poverty rate, to 24 percent."

Note that the U.S. poverty line is startlingly low. Various assessments suggested that the smallest amount needed by a family of four to purchase basic necessities in 1991 was 155 percent of the official poverty line. John Schwartz and Thomas Volgy wrote in the Feb. 15, 1993, issue of *The Nation* magazine:

> Since the [census] bureau defines the [working poor] out of poverty, the domi-

nant image of the poor that remains is of people who are unemployed or on the welfare rolls. The real poverty line reveals the opposite: A majority of the poor among able-bodied, non-elderly heads of households normally work full time. The total number of adults who remain poor despite normally working full time is nearly 10 million more than double the number of adults on welfare. Two thirds of them are high school or college educated and half are over 33 [years old]. Poverty in the United States is a problem of low-wage jobs far more than it is of welfare dependency, lack of education, or work inexperience. Defining families who earn less than 155 percent of the official poverty line as poor would result in about one person in every four being considered poor in the United States.

**Example:** The U.S. government rarely collects health data broken down by social class. In 1986, when it did this for heart and cerebrovascular disease, it found enormous gaps. As Vicente Navarro wrote in the April 15, 1991, issue of *The Nation:*

> The death rate from heart disease, for example, was 2.3 times higher among unskilled blue-collar operators than among managers and professionals. By contrast, the mortality rate from heart disease in 1986 for blacks was 1.3 times higher than for whites. … The way in which statistics are kept does not help to make white and black workers aware of the commonality of their predicament.

### Goal 4: Understanding the Politics of Knowledge

There are many aspects of the politics of knowledge that are integrated into this curriculum. Some involve reconsidering what counts as mathematical knowledge and re-presenting an accurate picture of the contributions of all the world's peoples to the development of mathematical knowledge. Others involve how mathematical knowledge is learned in schools. British educator Richard Winter, for example, theorizes that the problems so many encounter in understanding mathematics are not due to the discipline's "difficult abstractions," but due to the cultural form in which mathematics is presented. Likewise, Holly Sklar cites a U.S. study that recorded the differential treatment of black and white students in math classes:

> Sixty-six student-teachers were told to teach a math concept to four pupils—two white and two black. All the pupils were of equal, average intelligence. The student-teachers were told that in each set of four, one white and one black student were intellectually gifted and the others were labeled as average. The student teachers were monitored through a one-way mirror to see how they reinforced their students' efforts. The "superior" white pupils received two positive reinforcements for every negative one. The "average" white students received one positive reinforcement for every negative reinforcement. The "average" black students received 1.5 negative reinforcements, while the "superior" black students received one positive response for every 3.5 negative ones.

Class discussion of the above study can lead to examining these ratios and is assisted by creating matrix charts to visualize the data more clearly. Discussion also can involve students in reflecting on topics in mathematics education. This is another example of breaking down the dichotomy between learning and teaching.

Underlying all these issues are more general concerns that should form the foundation of all learning, concerns about what counts as knowledge and why. I think that one of the most significant contributions of Freire to the development of a critical literacy is the idea that "our task is not to teach students to think—they can already think—but to exchange our ways of thinking with each other and look together for better ways of approaching the decodification of an object." This idea is critically important because it implies a fundamentally different set of assumptions about people, pedagogy, and knowledge-creation.

Because some people in the United States, for example, need to learn to write in "standard" English, it does not follow that they cannot express very complex analyses of social, political, economic, ethical, and other issues. And many people with an excellent grasp of reading, writing, and mathematics skills need to learn about the world, about philosophy, about psychology, about justice, and about many other areas in order to deepen their understandings.

In a non-trivial way, we can learn a great deal from intellectual diversity. Most of the burning social, political, economic, and ethical questions of our time remain unanswered. In the United States we live in a society of enormous wealth, yet we have significant hunger and homelessness; although we have engaged in medical and scientific research for scores of years, we are not much closer to changing the prognosis for most cancers. Certainly we can learn from the perspectives and philosophies of people whose knowledge has developed in a variety of intellectual and experiential conditions.

Freire reminds us that currently "the intellectual activity of those without power is always labeled non-intellectual." When we see this as a political situation, as part of our "regime of truth," we can realize that all people have knowledge, all people are continually creating knowledge and doing intellectual work, and all of us have a lot to learn. ☐

_____

This article was adapted from a version that originally appeared in *Beyond Heroes and Holidays* (see Resources, page 267). For a listing of all citations and references included in the original article, see www.rethinkingschools.org/math.

# THE WAR IN IRAQ

## HOW MUCH DOES IT COST?

BY BOB PETERSON

Estimating the cost of the war in Iraq is like shooting at a moving target. The costs continue daily in amounts that many people cannot even imagine. For example, according to the website of the National Priorities Project, www.costofwar.com, by the end of January 2013, the cost of the war was over $810 billion. This is in addition to the $530 billion military budget that the U.S. Congress appropriated for fiscal year 2012.

There is a great deal of debate about how much should be spent on the military so that the United States is well-defended. Some people argue that if money spent on the military was instead spent on combating U.S. and global social and economic problems, we'd all be much safer.

Dwight D. Eisenhower, Commander of Allied Forces in Europe during World War II and the 34th president of the United States (and a Republican), said on April 16, 1953: "Every gun that is made, every warship launched, every rocket fired, signifies in the final sense a theft from those who hunger and are not fed, those who are cold and are not clothed."

### TOP GLOBAL PROBLEMS AND THE MONEY NEEDED TO SOLVE THEM

In 2000, the United Nations Millennium Summit brought together the largest gathering of world leaders in history to address global poverty. They agreed to eight Millennium Development Goals, and a time frame from 2000 to 2015 to reach these goals. The estimated cost for achieving these goals is $120–$190 billion per year.

Goal 1: Cut extreme poverty and hunger in half.
Goal 2: Achieve universal primary education.
Goal 3: Promote gender equality and empower women.
Goal 4: Reduce child mortality.
Goal 5: Improve maternal health.
Goal 6: Combat HIV/AIDS, malaria, and other diseases.
Goal 7: Ensure environmental sustainability.
Goal 8: Develop a global partnership for development.

continued on next page

## DO THE MATH AND QUESTIONS FOR DISCUSSION

1.   Assume you are able to set a $1 bill down on your desk every second without stopping. First have each student in your group record on paper your estimates of how long it would take to set down $1 million, $1 billion, and $156 billion. After recording your guesses, use a calculator to figure out how long each would take. Compare your estimates to your calculated answer.

2.   Using the www.costofwar.com website, estimate how U.S. taxpayers' money is spent on the war each second, each minute, each hour, each day, and each week.

3.   Look at the estimate of how much it would cost to reach the Millennium Development Goals set by the United Nations. How does the total cost compare to the cost of the war in Iraq?

4.   A full scholarship to an out-of-state, four-year, major university like the University of Wisconsin–Madison—including dorm fees, food, books, tuition, and other fees—is worth about $41,000 a year. Calculate how many high school seniors could receive a fully paid four-year college education if the monies spent on the Iraq war—as of January 2013—had been set aside for college scholarships instead.

5.   A group called the National Priorities Project has estimated the amount of taxes that people in different states and cities have paid to fund the war in Iraq. Go to their website, www.nationalpriorities.org, and find your state and/ or city. Using this number, calculate how many additional teachers and nurses could be hired in your community, assuming that each of them made the average national salary for their profession. In 2010-11 the national average teacher salary was $56,069 and the national average nurse salary in 2011 was $67,800.

6.   Re-read the above quotation from former President Eisenhower. What do you think about the quotation and how does it relate to our current political situation?

7.   Develop a short presentation for the class explaining what you learned and what you think about these matters. ■

# CHAPTER SIX

# Race, Retrenchment, and the Reform of School Mathematics

## BY WILLIAM F. TATE

The teaching of mathematics needs to be connected to the lives and experiences of African American students to enable them to fully take part in our democracy. Traditionally, schools have not provided African American students opportunities to do so.

More than 60 years ago, historian Carter Woodson described this dilemma:

> And even in the certitude of science or mathematics it has been unfortunate that the approach to the Negro has been borrowed from a "foreign" method. For example, the teaching of arithmetic in the 5th grade in a backward county in Mississippi should mean one thing in the Negro school and a decidedly different thing in the white school. The Negro children, as a rule, come from the homes of tenants and peons who live to migrate annually from plantation to plantation, looking for light which they have never seen. The children from the

homes of white planters and merchants live permanently in the midst of calculations, family budget, and the like, which enable them sometimes to learn more by contact than the Negro can acquire in school. Instead of teaching such Negro children less arithmetic, they should be taught much more of it than the white children, for the latter attend a graded school consolidated by free transportation when the Negroes go to one-room rented hovels to be taught without equipment and by incompetent teachers educated scarcely beyond the 8th grade.

One important implication of Woodson's argument is that mathematics instruction that is built on a student's life experience provides two mathematics learning environments: within the school and outside the school. Unfortunately, the disciplines that undergird mathematics education—mathematics and psychology—place great stress on objectivity and neutrality. As a result, school mathematics has been tacitly accepted as a colorblind discipline. Thus very little consideration is given to the cultural appropriateness of mathematics pedagogy.

## The Current State of Affairs

In recent years, mathematics textbooks have included pictures of African Americans, and some mathematics textbooks provide information about Africans and African Americans who have contributed to the development of the discipline of mathematics. These efforts represent progress and should be encouraged. Yet I doubt seriously if these efforts will prove sufficient to enfranchise African American students in mathematics. I contend that connecting the pedagogy of mathematics to the lived realities of African American students is essential to creating equitable conditions in mathematics education.

What type of pedagogy must African American students negotiate to be successful in school mathematics? Conventional mathematics pedagogy emphasizes whole-class instruction, with teachers modeling a method of solving a problem and students listening to the explanation. This is typically followed by having the students work alone on a set of problems from a textbook or worksheet. The goal of this teacher-directed model of instruction is for students to produce correct responses to a narrowly prescribed problem. This type of mathematics pedagogy is consistent with several studies of mathematics instruction conducted in the 1970s.

Unfortunately, this conventional mathematics pedagogy is exactly the kind of "foreign method" of teaching described by Woodson. Few if any attempts are made to build on the thinking and experiences of African American students. One important purpose of mathematics education is to prepare students to incorporate mathematical reasoning and communication into their everyday lives. However, conventional pedagogy has often persuaded students—particularly African American students—to consider school mathematics as a subject divorced from their everyday experiences and from their attempts to make sense of their world.

Today, the "foreign" pedagogy delineated by Woodson continues in many different forms. For example, pedagogy for African American students is hampered by the following conditions of their schooling:

- Persistent tracking.
- Less access than other students to the best-qualified teachers of mathematics.
- Fewer opportunities to use technology in school mathematics.

Moreover, as the proportion of African American students in a school increases, the relative proportion of college-preparatory or advanced sections of mathematics decreases.

What is the effect of the foreign pedagogy on African American students' thinking and achievement in mathematics? Data from the National Assessment of Educational Progress (NAEP) indicate that African Americans, across the grade levels tested, scored significantly better on portions of the test related to factual knowledge and basic computational skills than they did in previous years. However, no growth was evident on portions of the test that assessed more advanced levels of mathematical reasoning. Should we be surprised at these findings? Walter Secada commented on the consistency of the NAEP findings with the pedagogical focus of the previous decade:

> Indeed, if we agree that the 1970s and 1980s were times when educational policy was predicated on mastery of basic skills, we could read these data as telling a success story: Insofar as we had set a national goal—the acquisition of basic skills—we moved in the direction of its attainment. African Americans did become proficient in their mastery of low-level basic skills. Alternately, we might read this as a story of incomplete success: Insofar as basic computation skills are deemed insufficient for the knowledge and mastery of mathematics, these data document how much father we have to go.

## Race, Reform, and Retrenchment

What role should the reform of school mathematics play in changing this story of incomplete success? Reform documents published by the National Council of Teachers of Mathematics have called for a new vision of mathematics pedagogy. For example, the publication *Professional Standards for Teaching Mathematic*s recommends that all mathematics teachers know "how students' linguistic, ethnic, racial, gender, and socioeconomic background influence their learning of mathematics," and "the role of mathematics in society and culture, the contribution of various cultures to the advancement of mathematics, and the relationship of school mathematics to other subjects and realistic application."

Despite such recommendations, mathematics pedagogy built on the lives and experiences of African American students must overcome many formidable barriers if it is to make its way into our classrooms. Many scholars argue that the curriculum and pedagogy of mathematics have been, and continue to be, connected to Eurocentric precepts that exclude the African American experience. For example, George Joseph states:

> The progress of Europe and its cultural dependencies [including the United States of America] during the last 400 years is perceived by many as inextricably—and even causally—linked with the rapid growth of science and technology. So in the minds of many, scientific progress becomes a uniquely European phenomenon, to be emulated only by following the European path of social and scientific development.

Joseph's remarks are germane to the current debate involving equity in the reform of school mathematics. Rather than address the problem of how to connect mathematics to the lives of African American students, many advocates of reform have argued that African American students warrant better treatment in mathematics classrooms on the grounds of national economic interest and global competition. For example, a report from the Carnegie Commission on Science, Technology, and Government states:

> The national interest is strongly bound up in the ability of Americans to compete technologically. This requires not only an adequate supply of scientific and technical professionals but a workforce able to solve problems and use the tools of a knowledge-intensive economy. All young people, including the non-college bound, the disadvantaged, and young women must be given the opportunity to become competent in science and mathematics.

Arguments based on the need to prepare workers for life in a global economy, while persuasive to the public, generally overlook the problems facing African American students in many school settings. Michael Apple suggests that these appeals to economic self-interest are attempts to persuade educators to compromise their beliefs about individual human rights because of the current fierce global competition for economic prosperity. The tension between mathematics pedagogy that is connected to the experiences of African American students (an individual human right) and the national economic interest is real and can be a barrier preventing educators from seeking equity in mathematics education.

The documents that have spawned the movement to reform mathematics education have called for a new era of "mathematics for all." These documents have embraced the idea of improving the mathematics performance of African American students in our nation's urban centers, but achieving this objective will require a blueprint. As Larry Cuban warned, "Unless policymakers and practitioners begin to consider how problems involving schools are framed, they will continue to lunge for quick solutions without considering the fit between the solution and the problem."

The current reform movement in mathematics education has been framed within a discussion of national economic interests. This focus raises questions about the ability of current reforms to generate interest in pedagogical practices that

go beyond those intended to yield gains for the national economy. I contend that the educators and policymakers leading the mathematics reform movement should address two questions related to the problems facing African American students in

Susan Lina Ruggles

IN CONVENTIONAL MATH PEDAGOGY, LITTLE EFFORT IS MADE TO BUILD ON THE EXPERIENCES OF AFRICAN AMERICAN STUDENTS.

mathematics classrooms across the United States. First, should school districts support (and teachers provide) mathematics teaching built on the experiences and lives of African American students? And second, should the focus of mathematics teaching be to prepare African American students to participate in the national economy or in our democracy? Both questions are extremely complex and probably do not have one right answer. However, I will devote the remainder of my discussion to examples that help illuminate the urgency of these questions.

### "Centricity" and Pedagogy

I have argued that one barrier to an equitable mathematics education for African American students is the failure to "center" them in the process of knowledge acquisition and to build

## "Almost all the experiences discussed in American classrooms are approached from the standpoint of white perspectives."

on their cultural and community experiences. Molefi Asante defines the concept of centricity as "a perspective that involves locating students within the context of their own cultural references so that they can relate socially and psychologically to other cultural perspectives." The concept of centricity can be applied to any culture. "For the white students in America," Asante continues, "this is easy, because almost all the experiences discussed in American classrooms are approached from the standpoint of white perspectives and history."

The idea that mathematics pedagogy should be built on the experiences of the student is undergoing a resurgence in the current reform movement. Yet very little has been written about building a mathematics program centered on the thinking and experiences of African American children. I offer two examples of the influence of mathematics pedagogy that has not been constructed to center the African American child.

The first example is based on an experience with a student teacher at my university. She provided her class with the following problem: Joe has five pumpkin pies. Karen has six pumpkin pies. How many pumpkin pies do Karen and Joe have all together?

This problem was written on the chalkboard when I entered the classroom. Five children were seated in the room—one African American and four whites. This was a 2nd-grade class at the beginning of a school year. These five students were provided separate mathematics instruction so that they could receive individualized attention to help improve their performance. The four white children were busy using manipulatives and appeared excited about the process of solving the problem. On the other hand, the African American child was not engaged in any outward activity. He was very quiet and appeared uninterested.

I asked the student teacher about the African American student's behavior. She replied, "Mark does not like math. I don't understand why, he's not a bad kid." Curious, I inquired about the discussion that led to the pumpkin pie math problem. The student teacher responded, "I used pumpkin pie as the object to be added because Thanksgiving is in two weeks." She went on to say that she wanted to motivate the students to solve a problem related to their holiday experience. In essence, she was trying to center the students in the process of knowledge acquisition. She was using a cultural construct—Thanksgiving pumpkin pie—as the means of centering her students. Yet her attempt to center the African American student was unsuccessful.

I asked the student teacher if she thought every family ate pumpkin pie on Thanksgiving. Her response: "Probably." I asked her to ask each of the five children in the class. She discovered that pumpkin pie was indeed a Thanksgiving ritual in the homes of the white children. For the African American child, however, sweet potato pie was the dessert of the day. Thus the background discussion that led to the problem was "foreign" to this student.

The student teacher had worked to provide the children in her class with what she thought was a problem centered in the context of their lives. Yet the problem reflected the default position of an idealized white middle-class reality. The mathematics problem and her pedagogy were unwittingly constructed to be the property of white children.

Understand my point here: I am not suggesting that all African American families celebrate Thanksgiving in the same way or that all white families do either. Rather, I contend that the default position of mathematics curriculum, assessment, and pedagogy is often more closely aligned with the idealized experience of the white middle class. Moreover, this reality is subtle and often difficult to diagnose. The diagnosis is difficult because, traditionally, mathematics has been viewed as neutral and objective.

## Implicit Assumptions

This leads to my second example: A group of teachers from an urban middle school discovered this very point when a large number of students at their predominantly African American school responded "strangely" to an assessment item on a districtwide mathematics test. The basic structure of the test item was as follows: It costs $1.50 each way to ride the bus between home and work. A weekly pass is $16. Which is the better deal, paying the daily fare or buying the weekly pass?

The district's test designers constructed the problem on the assumption that students who solved the problem correctly would choose to pay the daily fare. Implicit in the design of the test item is the notion that all people work five days a week. It is also assumed that the employee has only one job. Yet these assumptions are not consistent with the daily lives and realities of many African American students. Thus it should not have been shocking that a large percentage of the students in this particular middle school thought that buying the weekly pass was the better option.

When school officials questioned the students about their responses, they found that many of the students had centered themselves in

> "Grandma understood many things that are barely recognized in the wider educational world even today. For example, she realized that arithmetic is injurious to young minds and so, after I had learned my tables, she taught me algebra. … She thought that … drill was stultifying. The result was that I was not well-drilled in geography and spelling. But I learned to observe the world around me."
>
> **— Margaret Mead,**
> reflecting on her own education at home

the solution process. These students converted the "neutral" context of the problem into their own property. For instance, the students commented on the fact that more than one family member could use a weekly pass. They also mentioned the option of using the weekly pass on Saturdays and Sundays. In the families of many African American students, the financial providers hold several jobs, and work both on weekdays and on weekends. For these students, choosing the weekly pass is economically appropriate and mathematically logical.

Taking a "centric" perspective would enable mathematics teachers to expect their students' responses to be linked to the specific details of the students' lives. Unfortunately, African American students who center their lives and experiences within the process of acquiring knowledge about mathematics risk being put down for focusing on "extraneous matters." As a result, many African American students view mathematics as a subject appropriate only for white males, and they fail to see the relevance and usefulness of the discipline.

A closer look through a centric framework reveals important details about the Eurocentric terrain that the children in this urban middle school had to negotiate in order to claim ownership of the mathematics presented on their test. First, the students had to understand that the mode of response should reflect the idealized experiences of a traditional white middle-class family. Second, the students had to understand the mathematics involved in solving the test item using the white middle class as a frame to guide their problem-solving. This dual consciousness was required if the students were to "succeed" according to the Eurocentric assumptions of the test designers.

## Dominant Culture

Is the first level of consciousness fair or empowering to African American students whose experiences are not those of the white middle class? William Harvey provides what I consider an appropriate response to this question.

> The advocacy of the values and lifestyles of the dominant white culture is not and cannot be psychologically beneficial to blacks. The dominant cultural value system, and that of the school, admonishes black students for being what they are, while physical and social reality prevents them from being anything else.

In order to take advantage of the diversity in a single classroom, a mathematics teacher's pedagogy should try to provide students with opportunities to solve problems using their experiences. Of course, the mathematical processes used to solve problems should be consistent with the contexts of students' lives. Furthermore, mathematics teachers working from a centric perspective would endorse having their students solve the same problem from the perspective of different members of the class, school, and society. This approach is consistent with the methods of successful teachers of African American children. Gloria Ladson-Billings found that these teachers looked to connect students' knowledge of self with broader social and political realities.

## Conflict, Democracy, and Pedagogy

A second barrier to providing an equitable mathematics education for African American students is the failure to prepare them for the conflicts of democracy.

Traditionally, mathematics education has been connected to issues of national economic survival, rather than to the development of democratic citizenship through critical thinking in mathematics. The latter involves helping students engage in mathematical thinking that is connected to the students' social and political contexts. The intent of this pedagogy is twofold: First, students are being prepared to take an active role in our democracy; and second, students are provided important insight into, and understanding of, the role of mathematics within the democratic system of governance.

Until recently, embedding mathematics pedagogy within social and political contexts was not a serious consideration in mathematics education. The act of counting was viewed as a neutral exercise, unconnected to politics or society. Yet when do we ever count just for the sake of counting? Only in school do we count without a social purpose of some kind. Outside of school, mathematics is used to advance or block a particular agenda. However, mathematics curriculum and pedagogy rarely prepare the African American student to engage in authentic contexts of democracy.

There is a growing consensus among scholars that knowledge is constructed through the interaction of the mind and authentic experiences. For example, John Seely Brown and his colleagues have theorized that in-school knowledge is acquired by working alone, memorizing static rules for well-defined problem settings. Acquiring knowledge of mathematics in this type of setting makes transfer difficult. In contrast, out-of-school knowledge is obtained by working in social settings in which it is important to know the reason for a conflict, to solve ill-defined problems, and to construct personal (i.e., centric) meaning. The activity, context, and culture of the learning environment are reference points for the retrieval, interpretation, and use of mathematical knowledge. A child's construction of meaning is a result of social interactions—the negotiation of understanding that derives from engaging in real activities of the culture.

Perhaps an example of the work of a practicing teacher will illustrate how mathematics pedagogy can build on the personal and social perspectives. The "War Room" is occupied by Sandra Mason's students. The 10 students in her class have targeted 25 problems that they feel are negatively affirming their community, including the 13 liquor stores within a 1,000-foot radius of the school. Each member of the class has been greatly influenced by the presence of the liquor stores (i.e., the problem is centric). For example, one female student commented: "We pass by drunks asking us for money. We keep going but they harass us and tell us to come here."

The students have found that the disproportionate number of liquor stores in their community is a function of local legislation. To resolve the conflict between the local liquor laws and their negative experiences, the students have formulated and proposed mathematically based economic incentives to get liquor stores to relocate away from the school.

For Mason's students, mathematics is thus more than just numbers shorn of context (i.e.,

in-school knowledge). Instead, mathematics is embedded within the culture of their daily struggles. These students are learning to use mathematics to support their political positions. For example, the students in this class have mastered the relationship between media "sound bites" and mathematics, because Mason has prepared them to ask and answer several questions, using mathematics, that will be important in launching arguments for their position or countering the arguments of their opponents. For example: What methods of mathematical analysis will best support our position? What

Susan Lina Ruggles

WORKING TOGETHER IN SOCIAL SETTINGS TO INVESTIGATE REAL-LIFE PROBLEMS MOTIVATES STUDENTS TO ACQUIRE OUT-OF-SCHOOL KNOWLEDGE.

variables should be included in our analysis to strengthen our position? How can we minimize the influence of variables that may weaken our position? Will percentages or raw numbers make a more striking impression?

These questions are integral aspects of "mathematizing" in our society. Yet such questions are rarely found in the sources that guide the development of mathematics reform. An African American child will confront similar "mathematized" renditions of social reality every day. For instance, the battle over the appropriate use of racial categories in the U.S. census is still being waged. The struggle over the construction of racial categories for the census is one of the most technically complex and socially relevant debates of our time. When will African American and other children encounter such a debate in school mathematics? Will the traditional curriculum and pedagogy of mathematics prepare our children to ask the types of questions that are necessary for participation in a democracy?

## The use of math in our democratic society is almost always linked to an attempt to secure control of property.

### Mathematical Production and Conflict

I propose the need for a mathematics curriculum and pedagogy that recognize mathematical production as a function of conflict on two levels.

First, like all communities, the mathematics community is governed by stated rules and a covert value system. Within this community, individuals and groups of mathematicians have a significant history of struggle over what is deemed appropriate knowledge for the discipline. The many other issues over which battles have raged within this community include the interpretation of data, credit for discovering ideas, and censorship. The very nature of mathematical production involves conflict. Yet traditional mathematics classrooms are structured to rank students' understanding of a body of static ideas and procedures. Hidden from the students is the role of conflict in advancing mathematical thought.

Second, the use of mathematics in our democratic society is almost always linked to an attempt by one group or individual to secure control of property. This means a social,

political, and/or economic conflict will necessarily ensue.

In economic terms, there are two conditions at work when issues are mathematized. First, situations are mathematized in order to maximize the return on the information that is being analyzed. Thus mathematics pedagogy should seek to prepare students to answer the following question: What agenda is being maximized by a given example of mathematizing?

For example, hospitals have experimented with computer systems that can calculate the probability of surviving an illness, based on specific variables about the patient. This development provides teachers with an opportunity to explore with their students a mathematized situation. For instance, decisions to place patients on life-support systems could be influenced by the mathematical models used by the computer. Mathematizing the decision to place a patient on life support—under the banner of optimizing system use, minimizing hospital cost, and maximizing profit—could result in discriminatory practices. Adopting the personal (centric) and social (conflict) perspectives would lead mathematics teachers to explore a variety of questions with their classes. What variables are included in this model? How do variables such as gender, family history, and income influence a patient's chances of receiving life-support services? Does being an African American male lower a patient's chances of receiving life-support services, since, comparatively speaking, African American males have a shorter life expectancy? In essence, does the computerized mathematical model reinforce existing survival rates?

The second reason situations are mathematized is to minimize the risk of successful challenges to the decision-making process.

How many students leave school with enough knowledge and practice to challenge the use of mathematics in society? Thus far, it has been the case that those few who have learned to use mathematics as a tool to guide their own decision-making have stifled the voices of the large segment of the population that does not know how to counter with their own mathematically based arguments—a disproportionate number of whom are African Americans.

## Learning to Analyze Mathematized Situations

If students learn in school to analyze and critique mathematized situations, such as the hospital computer system described above, they will be prepared for public discussions about the development and implementation of the new models that are used in social decision-making. Such preparation is radically different from merely preparing students to add, subtract, multiply, and divide accurately. Our highly technological society requires that all, not just African American, students be prepared to use mathematics to defend their rights. The curriculum and pedagogy of mathematics should support this objective.

I have argued that African American students should be provided a mathematics pedagogy that is built on their cultural experiences. Furthermore, I contend that this mathematics pedagogy should prepare African Americans and others to participate in our democracy. Traditionally, these two recommendations have not been part of the "mainstream" discourse on school mathematics. More recently, standards for school mathematics reform have called for pedagogy that recognizes these objectives. However, research suggests that many educators will most likely implement only those mathematics standards that are consistent with their own beliefs about appropriate mathematics pedagogy.

Will teachers and other educators view as important the dual need to build on students' personal experiences and to empower them to function in a democracy? Will they move to change current practice? Woodson provides an appropriate ending to this discussion:

> In the first place, we must bear in mind that the Negro has never been educated. He has merely been informed about things which he has not been permitted to do. The Negroes have been shoved out of the regular schools through the rear door into the obscurity of the backyard and told to imitate others whom they see from afar, or they have been permitted in some places to come into the public schools to see how others educate themselves. The program for the uplift of the Negro in this country must be based upon the scientific study of the Negro from within to develop in him the power to do for himself.

The prospect of a new beginning for mathematics education rests with the ability of mathematics teachers to provide pedagogy that builds and expands on the thinking and experiences of African American students. Moreover, this pedagogy should focus on preparing these students to function within our democracy. ☐

This article is adapted from a version that originally appeared in *Phi Delta Kappan* magazine.

# ENVIRONMENTAL HAZARDS

## IS ENVIRONMENTAL RACISM REAL?

BY LARRY MILLER

This activity makes use of an interactive online map maintained by the U.S. Environmental Protection Agency (EPA), which shows the location of environmental hazards in considerable detail.

The purpose of this activity is to investigate whether environmental racism is real.

### PROCEDURE

1. With students, identify the zip codes in the area in which they live (or in some urban area).

2. Go online to the interactive map at http://www.epa.gov/ myenvironment/. Type in a zip code, city, or state, and click the "Go" button to the right. A selection of options will appear.

3. Under MyMaps, click on the "More Maps" link, to take you to the map of your chosen area. This map will include a "Map Contents" box. Click on "LAND," and you will see a selection of choices including "Superfund sites," "hazardous waste," "toxic releases," and "brownfield properties."

4. As you click on the check boxes next to each of these, the map will be redrawn map showing all the identified EPA sites in that area.

5. Have students follow the procedure for steps 6 through 10 below.

6. Count the number of city blocks shown on the map from east to west, and divide that number by 12 to determine the length in miles. Do the same from north to south.

7. Multiply the length by the width in miles to get the size of the zip code area in square miles.

8. Count the number of identified EPA sites (all colors) on the map.

9. Put the number of sites in ratio to the square miles. This is the density (number of sites per square mile.)

10. Divide the number of sites in a zip code by the square miles. This will give density per square mile. This ratio is needed to compare zip code areas because of their different sizes.

continued on next page

11.  Repeat steps 2 through 10 using the zip code of an adjacent suburban area.

12.  Compare the density of hazardous sites in students' neighborhoods or the urban area to the density in the suburbs. (In Milwaukee, students see a huge disproportion when they compare the ratios found in the city and the suburbs.)

13.  Have students read this definition of environmental racism:

> Environmental racism is racial and class discrimination in environmental policymaking and the enforcement of regulations and laws; it is the deliberate targeting of people of color and their communities for toxic and hazardous waste facilities and the sanctioning of the life-threatening presence of poisons and pollutants in poor communities and communities of color.

14.  Students then answer and discuss the question: "Is environmental racism real?"

15.  Have students visit the Voices from the Grassroots website, www.ejrc.cau.edu/voicesfromthegrassroots.htm, to read reports from around the country about the environmental justice movement and report on their findings to the rest of the class. ■

---

For a detailed description of how one class took action based on this lesson, see www.rethinkingschools.org/math.

STUDENTS FROM LARRY MILLER'S CLASS HELPED EXPOSE A TOXIC MESS IN THEIR NEIGHBORHOOD.

# CHAPTER SEVEN

# Adventures of a Beginning Teacher with Social Justice Mathematics

### BY LIZ TREXLER

It was during my first semester of student teaching at King College Prep High School (92 percent African American) in Chicago Public Schools that I was inspired to create my first social justice mathematics lesson. Having looked through student interest surveys, I discovered that many of my students suffered from asthma—a common ailment stemming from genetics and poor air quality. In my Algebra I class, we connected proportional reasoning to patterns and rates of children with asthma in Chicago by comparing data from particular neighborhoods.

After the lesson, I wondered how effective mathematics in this context was in developing a deeper mathematical understanding. Sure, my students appeared more engaged than if we were just doing traditional math, but how was I to be sure that their proportional reasoning benefited from this? During the lesson, several students asked, "Isn't this math class?" or "Why are we talking about asthma?" When they posed this question in other classes, in looking at charts or making sense of problems in science and social studies, did they recognize that they were reasoning mathematically? This led me to the question: Do students perceive social justice mathematics as being relevant to their daily lives?

I turned to the most valuable source I had in my classroom: my students. I formed a focus group of 15 students to get a sense of what they considered meaningful mathematics. I asked what gives them the most problems in mathematics, what makes a problem "real-world," and to describe who and what a mathematician is. Every response to this last question—someone who knows math inside and out, someone who majors in math, a person who understands concepts and formulas and when to apply them—involved being somewhat of a "genius," as the students put it. When I asked whether they saw themselves as mathematicians, most responded that they were not confident and/or knowledgeable enough to earn this title. I wanted to inspire them to change this perception. To me, a mathematician is a person who can not only apply mathematical concepts, but can also use them to make sense of the social, political, and economic world in which they live.

Being able to use a concept on a daily basis seemed to be valuable to them. One student said, "If you talked about texting, I guarantee you everyone in this classroom would listen. We love texting." Although I could easily incorporate this into the curriculum, I asked myself, "Would I be integrating this because it would provide a rich mathematical experience or to please my students?" This topic may be relevant to them, but they would not necessarily gain a better understanding of an algebra concept or an injustice in their world merely because I mentioned texting. I would need to situate the mathematics in meaningful contexts for my students to truly understand how math could reveal things about their world.

When I asked for characteristics of a real-world problem, they responded that it was something that related to students. I asked a follow-up question: Is it possible to create one question that would relate and resonate with each and every student in the classroom? They all told me probably not. This was something that I never thought about. Many application problems can indeed be real-world, but perhaps not for every learner.

I also had students evaluate three math problems (below)—rating how real-world each problem was, how much the problem interested them, and how the problem helped them to make better sense of the world they lived in.

1. Several students go to the store. They buy 12 candy bars that all cost the same and spend $5.16. How much does each candy bar cost?
2. A factory worker in Honduras works 12 hours a day and earns $5.16 in daily wages. How much does he/she earn per hour?
3. It costs $1.50 each way to ride the bus between home and work. A weekly pass is $16. Which is a better deal, paying the daily fare or buying the weekly pass?

The results in comparing problem #1 and #2 were interesting. Although more students felt that #2 better informed them about other situations in the world, they felt that problem #1 helped them make better sense of their *own* world. The majority of the group also felt that problem #1 was more effective in getting the math concept across. Clearly, students would encounter a situation such as buying candy more on a day-to-day basis than calculating wages of children in poor countries, so they evidently felt better connected to problem #1. However, would I have gotten the same response if I had changed #2 to read, "In your first after-school job, you work 12 hours a day and earn $5.16 in daily wages"? Would putting them in the context of the problem make them evaluate it any differently?

Students rated number #3 the most real-world by far. As they worked, one student asked, "Does a week mean five days or seven days?" Another asked if riders were making transfers on the bus. Yet another asked if they should count Sunday, since many people didn't work on Sundays. Although written for a standardized test, this question had a specific answer in mind; however, this demonstrates there are, in fact, several correct answers. This question illuminated that students take what they need from the outside world to make sense of real-world situations in the classroom.

The asthma lesson and focus group shaped my view of effective social justice math lessons in two ways. First, I gained a better understanding of what my students thought of as meaningful mathematics. From a survey given to the entire class, I found it fascinating that although students felt that the asthma lesson was less applicable to their everyday lives, overall, they found the activities engaging. However, students in the focus group said that they could not only apply proportional reasoning better to other contexts beyond asthma, but they saw also how this type of math could unveil characteristics about their own neighborhoods. This indirectly led to my second finding: that the math itself should genuinely highlight—and serve as a tool to understand—issues of social justice that students care about. Many of the problems I created for the asthma lesson were a bit of a stretch and not genuine problems they would encounter outside

of the classroom. Although I brought in topics of social justice to accompany the math, the math itself did not always highlight these issues.

I decided that I would make another attempt at creating a lesson involving a social justice topic using my newly acquired understandings. After my year of student teaching in Chicago, I moved to Denver to officially begin my teaching career in a small district where the student population is 80 percent Latina/o and 85 percent of students qualify for free or reduced lunch. Although the socioeconomic status of my students in Denver was similar to my students in Chicago, the surrounding neighborhood and community created some apparent differences in academic culture. One main difference between my students in Chicago and those in Denver was the percentage of students who pursued postsecondary options. Of the graduating class of 2011 at my high school in Denver, only 35 percent of students continued their education beyond high school. At King College Prep, 85 percent of students pursued postsecondary opportunities.

Since I have been teaching in Denver, there has been a big push to increase awareness of postsecondary opportunities for students in

## UNDERGRADUATE RESIDENT TUITION AT COLORADO STATE UNIVERSITY
### (BASED ON 30 CREDIT HOURS)

| 2005–06 | 2006–07 | 2007–08 | 2008–09 | 2009–10 | 2010–11 | 2011–12 | 2012–13 |
|---------|---------|---------|---------|---------|---------|---------|---------|
| $3,381 | $3,466 | $4,040 | $4,424 | $4,822 | $5,256 | $6,307 | $6,875 |

## UNDERGRADUATE NONRESIDENT TUITION AT COLORADO STATE UNIVERSITY
### (BASED ON 30 CREDIT HOURS)

| 2005–06 | 2006–07 | 2007–08 | 2008–09 | 2009–10 | 2010–11 | 2011–12 | 2012–13 |
|---------|---------|---------|---------|---------|---------|---------|---------|
| $14,343 | $14,994 | $17,480 | $20,140 | $20,744 | $21,366 | $22,007 | $22,667 |

hopes of increasing the percentage of students who pursue these options. In my linear functions unit for my Algebra II class, I thought students could benefit from examining and comparing tuition rates between colleges, as well as trends over time, since the cost of attending college has steadily increased. I told my students that we would use linear regression to analyze tuition trends at Colorado State University, the closest state university to our high school, as shown on page 56.

I asked students to write down one question that this data evoked. Most of their questions asked, "How much will tuition be when I am a freshman in college?" This was obviously relevant and necessary if they were to financially plan for college. They could also investigate this using mathematics. However, several students thought more in-depth, asking, "Why does tuition increase more in certain years than others?" and, "Why is nonresident tuition so much more?" These questions didn't necessarily address the math concepts we were studying, but were helpful in understanding the problem context.

My students already knew how to use a graphing calculator to input data and find the line of best fit. However, I wanted to use the characteristics and information provided by the equation of the line of best fit, in slope-intercept form, to compare the scenarios.

Using linear regression on the data above, we obtained the following equations in slope-intercept form (where x is the number of years since 2005–06 and y is the cost of tuition for that year):

Resident Tuition:
$$y = 508.44x + 3{,}041.83$$

Nonresident Tuition:
$$y = 1{,}257.08x + 14{,}817.83$$

Several questions immediately arise. What do these numbers mean in the context of the problem? In order for students to analyze and make inferences from the data, they first needed to understand what the slope and y-intercept represented in the real-world context beyond the equation or the graph. From the equation, we can see that resident tuition in 2005–06 (when x = 0) should be $3,041.83. However, from the table, we see that tuition was actually $3,381. This may cause students to question, "Why is it different? What do the intercepts tell us about approximation and lines of best fit?" From this data alone, students can use math to inform their own thinking and ask more personal questions, such as, "Will this affect me?" or, "How much more expensive will tuition be when I attend college in a few years?"

Another helpful question in analyzing the data would be: "If we looked at a subset of the data, how would these equations change?" For example, if we examined only tuition for

> "Is this fair that they have to pay more to get the same education just because they don't have a piece of paper?"

2009–2013, how would our slopes and intercepts change? Would this be a better model to predict the cost of tuition in the immediate future? In answering these questions, students not only demonstrate their understanding of what creates an equation, but the math can also become a springboard for further investigation.

One key aspect that highlighted an injustice specific to my students was the difference in the slope between a resident and nonresident student. For a student attending Colorado State University as a resident, one could predict tuition to increase by $508.44 each year. For a student attending as a nonresident, tuition was predicted to increase by $1,257.08 each year. The response to the question, "Who pays resident vs. nonresident tuition?" is fairly simple for most: Resident means you reside in that state and nonresident means you do not.

However, many of my students are undocumented, which meant that if they wanted to attend Colorado State University or any other public university in Colorado, they would pay nonresident tuition. One undocumented student realized that over the course of four years, he would pay nearly $63,000 more than his friend who is documented. This caused another student to ask, "Is this fair that they have to pay more to get the same education just because they don't have a piece of paper?"

These low-income students who already have difficulty affording college regardless of immigration status saw how the trends in tuition prices disproportionately affect particular populations, making it harder for undocumented students to pursue the same opportunities as their documented peers. By the end of the lesson, I felt that my students had not only explored concepts of linear regression, but they also recognized how useful this tool could be to analyze their future. The comments and discussions I heard between students illuminated an injustice in their own community.

Upon reflecting on my initial lessons from student teaching, I realized that I was putting too much effort into creating an ideal, super lesson that would take a topic of social justice and then integrate the math into it. The mathematics in a situation should highlight questions and concepts of social justice to create a cohesive and authentic opportunity to investigate. That is what I tried to do with the tuition context.

According to Bob Peterson in the first edition of *Rethinking Mathematics*, if students do not see how math applies to their lives, they are "robbed of an important tool to help them fully participate in society." Seeing math as a means of investigation is one way that students can begin to view math as the invaluable tool that Peterson describes. It can also serve as an entry point to more involved conversations. If we are to encourage students to use math to make sense of the world around them, we need to provide them with mathematical contexts that give them the experience in doing so.

I hope that motivating students to use mathematics as a tool rather than just a subject they have to take can change the question of, "When will I use this?" to, "How can I use mathematics to make sense of the reality in which I live?" ☐

## PART 2

David McLimans

# Infusing Social Justice into Math Classes

# CHAPTER EIGHT

# "Home Buying While Brown or Black"

## Teaching Mathematics for Racial Justice

### BY ERIC (RICO) GUTSTEIN

So one question leads to another question, and then you have to answer four more, and those four questions lead to eight more questions. So I think that [racial disparity in mortgage lending] is not racism, but that leads me to the conclusion that if it was not racism, then why do they pay more money to whites? Is that racism?

**— Vanessa, 7th grade**

Vanessa wrote these words on a mathematics project I taught to middle school students titled "Mortgage Loans—Is Racism a Factor?" To begin the project, I had the students read a *Chicago Tribune* article that analyzed mortgage rejection rates for African Americans, Latinos, and whites in 68 different metropolitan areas. The students mathematically analyzed data and wrote essays about whether racism was a factor, using data and arguments from the article. The mortgage project is an example of how I use mathematics to teach for racial justice.

A central part of teaching for social justice is to work for a society where racism is reduced and, eventually, eliminated. But racism is so deeply embedded in the history, consciousness, and fabric of U.S. life that to remove it will take long-term, concentrated efforts. We need to understand racism, its genesis and manifestations, and also build schools and classrooms that explicitly promote racial justice (and that are linked to anti-racist social movements on a broader scale).

I have occasionally taught middle school mathematics in a Chicago public school, located in a low-income, Mexican immigrant community, as part of my job as a university-based mathematics educator. I use ideas of teaching for social justice along with helping students develop mathematical power (being able to reason and communicate mathematically, develop their own mathematical thinking, and solve real-world problems in multiple and novel ways)—and pass the "gatekeeping" standardized tests.

Through studying these experiences and other learnings, I have come to certain ideas about teaching mathematics for racial justice. In my view, a racial justice curriculum should provide students opportunities to:

- Develop an understanding of the socio-political, cultural, economic, and historical dynamics of racism, along with their interconnections.
- Appreciate the complexity of different forms of racism (structural, institutional, individual).
- Develop and use analytical tools (such as data analysis, graphing, and mathematical modeling of social phenomena) to understand and dissect racism.
- Support views, develop coherent arguments, and engage in group discussions to develop individual and collective analyses.
- Develop an appreciation for multiple, alternative, and competing perspectives

while constructing their own independent knowledge.
- Become active, get involved in actual struggles when possible and appropriate, and use their analytical tools to cut through and confront myths and historical inaccuracies.

Another important issue for teachers is that classroom cultures supporting investigations of complex and emotional issues like racism have to be co-created with students over time. One cannot just "plop down" potentially volatile issues without creating conditions for students to take seriously their roles as learners and knowledge creators. Students need to be able to listen to others and find their own voices as well. In highly regulated urban schools, creating classroom norms that support this type of work can be challenging and takes time, because such a culture contradicts students' expectations of what teaching "should be"—and students are not always so quick to accept these new routines and new topics of discussion.

This last point is particularly important when white teachers are teaching about racism and for racial justice with students of color. As a white teacher of students of color, I consider the following questions: Why should students of color, who know firsthand the pain and horror of racism, necessarily share their thoughts, feelings, and ideas in a classroom with a white teacher? What difficulties exist when teaching about racism across color lines, given sociopolitical and cultural power differentials in a racist society?

Since my students have direct experience with racism, and my knowledge is more analytical, historical, and observational, how do we meet in mutual respect to deepen our collective understanding of the impact of racism and how to fight better against it? How do I better understand how my students perceive and experience racism and other forms of discrimination so that I can create opportunities for them to better comprehend the nature of these injustices—all while teaching mathematics in a high-stakes, urban school district focused on "accountability," regimentation, and discipline?

I have found it challenging, but possible, to try to enact the above principles in the classroom. What follows is a lesson in which I tried to put these ideas into practice.

### An Example of Racial Justice Curriculum

In December 2002, I gave my students the mortgage project. (Materials for this project are available online at www.rethinkingschools.org/math). This lesson was challenging because the article from the *Chicago Tribune* was confusing, the mathematics was complicated, and terms were used without definition. But it was clear that in Chicago, African Americans were rejected for mortgage loans five times as often as whites, and Latinos were rejected for mortgage loans three times as often as whites. This was true across all income levels, negating income alone as an explanation.

The project took almost three weeks of homework and in-class time. Students worked mainly in groups, although all wrote individual essays. I had most of the students rewrite their essays one or two times, usually because they did not adequately support arguments, and most groups reworked other parts of the project as well. We ended with two days of students reading their essays aloud and trying to arrive at individual and group understandings. Throughout the lesson, students were extremely engaged.

When I look at their projects, essays, and journals, several themes are apparent. First, many students learned that it was difficult to know

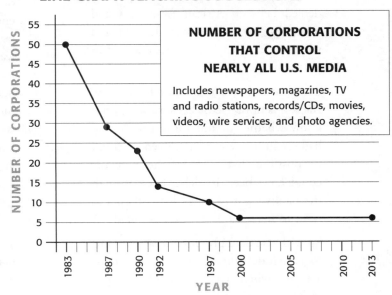

## CORPORATE CONTROL OF U.S. MEDIA

### LINE GRAPH TEACHING SUGGESTIONS

**NUMBER OF CORPORATIONS THAT CONTROL NEARLY ALL U.S. MEDIA**

Includes newspapers, magazines, TV and radio stations, records/CDs, movies, videos, wire services, and photo agencies.

Ask students to compute a best-fit equation for the line. What is the prediction for the future? What is the mathematical problem with extrapolating this line too far into the future? How would you get around that? What do you think is the effect of the media concentration? What does this make you think about watching the nightly news? But after the year 2000, the number of corporations controlling the media stabilized at six. If students *just* looked at 2000-2013, they would miss the change over time and could tell a different story. Some might argue that there's no problem with media concentration. Discuss with them how the different starting points of such a graph affect how we interpret reality.

if racism was a factor. Many students changed their minds, often more than once, after rethinking, further questioning and investigating, and listening to their peers' essays. I considered this important, because too often students (and adults) gravitate to simplistic "all-or-nothing" solutions and overlook real-world complexities. Most appeared comfortable with ambiguity and the realization that they needed more data to answer the question. For example, Jesse wrote:

[Racism] is a factor because white applicants, no matter what their income was, they were always denied less times than African Americans and Latinos. And it is also a factor because the ratio of applicants denied between African Americans and whites is 5:1 and between Latinos and whites is 3:1. That data shows that racism is a factor.

There are always two sides to a story. Racism is not a factor because we do not know whether or not those people had bad credit or were unemployed. It could be possible that a lot of those people could have been in debt. Even though the banks want to make loans they also want to make sure that they get paid.

So with the data provided it is very hard to conclude whether or not racism is a factor when it comes to obtaining a mortgage loan in the Chicago area.

Manny went back and forth several times; his final journal entry read:

My view is that racism is a factor. First I was unsure if it was or wasn't, so I did an essay saying it wasn't but at the end changed my mind that it is. I have learned some things about mortgages. I learned that blacks have a harder time than Latinos. It's basically from really light skin to dark. Blacks have five times more trouble to get a mortgage which is a lot more times compared to whites. Latinos don't have it as bad as blacks, but not as easy as whites. My only question is: Is racism also in job applications?

And Nilda wrote:

Was this project to confuse us and really make us think? Because that's what it did. After our last [whole-class] discussion on Friday, everyone was talking

**FRED WRIGHT**

about what we had discussed. In my first article I said that I thought racism was not a factor, after our second discussion I thought racism was a factor, but I think that we don't really know. Even though the rate for blacks was five times higher than whites in being rejected, that does not necessarily mean it is racism, it could be because of their debt, income. Or maybe it could be racism.

An interesting point Nilda and others made is that students continued the discussions outside of class, testifying to their engagement—a typical written comment was, "me and some of the girls have been talking about the conversation…"

In addition to discovering that "one question leads to another," the students also constructed alternate explanations through this process. Vanessa wondered if racism accounted for income disparity, Jesse pondered whether racism was a factor in job applications, and Nilda proposed debt and income as other possible causes of the rejection-rate disparity. Students did not base these ideas so much on the article, but rather on their own knowledge and experiences. Common explanations were "bad credit," "less education," "worse jobs," and "no

papers" (undocumented). But those explanations often led to other questions, as Carmen wrote:

> I do not think racism is a factor in mortgages. At first I thought it was but not until I got to question number seven, when I defended Bank One [they had to take the bank's position in this question]. I said that if they were to ask me if racism was a factor in mortgages I would say something like this, "Racism is totally not a factor in mortgages, in fact the reason why a lot of blacks and Latinos are being rejected is because they have less collateral." In case y'all don't know what collateral is I'll try explaining it. Collateral is something you have/own that can substitute [for] money (the loan) if you can't pay—somebody asked me why whites had more collateral and I said because they have been here (in U.S.) longer. Then that one person asked me why didn't blacks have more collateral if they've been here longer, and to that question I would say that since a long time ago whites owned lots of things and blacks were not able because they were slaves.

# "AND HOW DO YOU SPEND YOUR WAGES?"

## CARTOON TEACHING SUGGESTIONS

Show students the cartoon at left and ask, "What is the message of the cartoon?" Have students find the federal and state minimum wages and then calculate how much money a person who earns the minimum wage, working 40 hours per week, would have to put toward the basic necessities of food, rent, clothes, health care, transportation, etc.

Ask, "What measures could working people take in order to receive adequate wages? What role could math play in their efforts?"

Carmen laid bare some of the complexity but seemed to be unaware of it. She essentially named racism as a factor in African Americans lacking collateral (wealth), and hence their difficulties in securing mortgages, but did not make all the connections. A possible question to ask her might be, "Since blacks having less collateral is a factor in not getting mortgages—is racism a factor after all?" This theme, of one question leading to another, arose throughout the project, and I also promoted it by asking students to generate their own questions.

Despite my attempts, I felt that many students did not grasp the complicated sociohistorical reasons for the racial disparities in lending. Carmen came closest, but overall, I was unable to help them understand the relationships. Even though at the end of the project I gave students a handout comparing black and white income and wealth (assets) in the United States, we didn't spend enough time working through the meaning and situating it historically. It was the end of the project, right before winter break; students had been working hard and were tired, and I planned to return to it later in the year. Nevertheless, this was a weakness in the project.

During these discussions, the students' own stereotypes surfaced. Two students read aloud their essays, stating that African Americans had "dirty houses," and suggested that as a reason why they were rejected at higher rates. This sparked excellent discussions about the issues, and the general consensus emerged that one person of color cannot be seen as an example for the whole. Several students reacted, both verbally and in writing, and identified the stereotypes placed on themselves, as Latinos, as a point of unity. Laura wrote:

> That's a stereotype. I got kind of mad, because it's like people thinking Mexicans are lazy. There's this "famous" thing/picture of a Mexican guy with a sombrero hiding his face, his head on his knees, and sleeping. That's a stereotype also. It doesn't mean it has to be true.

That unity was important because it provided a basis for students to understand common oppression. This in turn can lead to students seeing themselves and others as mutual allies, creating conditions for them to act.

## Making Connections

A project like this represents one way to teach mathematics for racial justice. I think it's important that students began to go beneath the surface and grapple with the cultural, economic, and historical roots of racism. The questions they asked were attempts to make sense of complicated data and a complex social phenomenon (racism) that many adults have trouble with. They did not, for the most part, fully differentiate between individual racism (e.g., a white loan officer rejecting an applicant of color because of her race) and structural racism (e.g., why African Americans have less collateral). Nor was I able to help them understand the sociohistorical context. But I believe this awareness is part of a developmental process—after all, we cannot just "tell" people an analysis, they have to construct their own interpretations. And they did begin to question and make connections.

I wanted students to raise questions, publicly discuss, and seriously study these issues. The evidence was there that they achieved these goals, while learning mathematics and using mathematics as an analytical tool with which to understand racism—even if some of their questions remained unanswered. In addition, I wanted to learn about students' perspectives and life experiences in order to be clearer about my own role in the struggles for racial justice, and, to a certain degree, most were open and willing to share their lives.

Above all, I wanted the students to have an inquiring, critical perspective. As Tita wrote, "From this project I learned that you should question everything. Like that to have a better project you should question all your answers. That's what I did." □

All students' names have been changed.

# CHAPTER NINE

# Living Algebra, Living Wage

## 8th Graders Learn Some Real-world Math Lessons

BY JANA DEAN

I know people don't usually ask. But do you mind telling me how much you make an hour?"

I felt a little uneasy asking this question at the checkout stand at the big-box store next to the freeway. It's not something you usually talk about when exchanging money for groceries.

The checker was happy to answer. "Well, I started at $9.12," she said. "Last month I got a 22-cent raise. I thought it would make a big difference, but I can hardly tell. My paycheck's almost the same size." [Editors' note: This article was first published in *Rethinking Schools* in 2007, so specific dollar amounts may be out of date.]

I explained the reason for my question. "I teach math at the middle school down the street, and I want to use wages to teach them algebra."

"Well, you can tell your students there's no way to make it on less than $10 an hour," she said. "Costco's where they want to work. They start you at $10." I thanked her and paid for my groceries.

In using wages from our community to teach about linear relationships—mathematical relationships in which the rate of change is steady and graphs as a straight line—I had two goals.

First, I wanted my students to engage their skills in math class to inform an ethical stance on a social issue close to their own lives: working for a decent wage. They would calculate daily and monthly incomes in service sector jobs, discover hidden costs associated with being employed, research local housing prices, compare state and national minimum wages, and learn about the activism and organizing effort behind Washington state's highest-in-the-nation minimum wage. Ultimately, they would use their knowledge of economics and mathematics to develop a point of view about the minimum wage.

My second goal had to do with motivating students to stick with algebra. Giving all students access to higher math regardless of family background is one reason I teach middle school math. Socioeconomic status informs student experience both in and out of school. In my school, students are tracked starting in 8th grade, which is when algebra comes around. The placement of students tends to fall along class lines. A few weeks ago, when I asked how many of my advanced track students know someone who has taken out a payday loan, a few timid hands went up. Most didn't know what I was talking about. When I asked my other three classes the same question, they had plenty of experience to draw on: Almost everyone knew someone. Decent, middle-class paychecks keep most people from having to borrow at 375 percent annual interest to make it to payday. In using the topic of a living wage to teach algebra, I hoped to build a bridge between my students' lives and algebra.

### A Living Wage

Advocates for the working poor set living wages for U.S. communities by researching typical housing, childcare, food, medical, and transportation expenses in a given community. According to the Economic Policy Institute (EPI), the living wage for the Tumwater, Wash., area, which includes no "extras" such as new shoes, big-screen TVs, or birthday presents, is $39,000 for a family of three, or about $19 an hour full time. This is about twice the federal poverty level. Agencies use the poverty level to determine government assistance and as an economic indicator. While 12.6 percent of our state's population lives below the federal poverty level, 26.9 percent of all people live in households whose income is below a living wage, as determined by the EPI. Most of those households have heads of household who are employed. This mirrors the nation.

Thinking to launch our study with an engaging story, I read out loud from Barbara Ehrenreich's *Nickel and Dimed*. I chose a section in the middle of the chapter "Selling in Minnesota" in which Ehrenreich describes her struggle to pay for the clothes she needs for her new job at Walmart and her frustration at how tedious and difficult the work is. I chose it because I thought it illustrated the complexities and compromises that come with accepting a low-wage job.

My students didn't respond at all. Rather than the open-eyed engagement I'd expected, they zeroed in on the fine art of finding and tracing nicks and scratches in the surfaces of their desks. Either I'd completely missed the mark, or I'd struck too close to home.

### Tapping Students' Opinions

Luckily, I had a planning period before I was to teach this lesson to my next class. It was possible my students didn't need Barbara Ehrenreich. Maybe they had some personal experience I could draw on instead. To my next class, I read the following statements out loud:

1. Those who work should be paid.
2. No one who works full time should live in poverty.
3. Wages should be high enough to support a family on one income.
4. The legal minimum wage should be high enough to get by.

I then asked students to talk to their partner about each statement and to say whether they agreed or disagreed with it. Afterward, they wrote their responses in their notebooks.

The contrast with the previous class was stunning. As the room filled with their voices, I knew I'd hit the jackpot. They spoke from experience. They all connected work with income: Everyone thought that anyone who works should be paid. No one thought that people who work full time should have to live in poverty. They had mixed responses to my prompt about whether or not a family should be able to cover basic needs on one income, and revealed that the community norm is that at least two people in a household work.

Daryl's response was typical: "The legal minimum wage should be high enough to get by because if people aren't making enough working full time, they need to be paid more."

Stephanie's comment sounded personal, "People work hard for their money ... It's expensive to pay bills, buy food, pay rent and all the extras."

I knew I had students engaged: Ethical consideration of fair pay interested them. But I wasn't sure yet if the head of steam we'd gathered would carry into the hard work of learning algebra. Sadly, by middle school, students expect math class to be disconnected from their lives. This gives an extra boost to any lesson that occurs within math class that appears to contain no math. "Thank goodness," students seem to say to themselves. "Finally we're doing something that matters!"

So before the bell that day, I threw in some numbers. I wanted to take advantage of my students' interest in the social issue of fair pay and connect it right away with mathematics. I asked them to estimate the federal minimum wage. Most students guessed on the order of $10 an hour. When I told them it was $5.15 an hour, they were

shocked. Jay exclaimed, "Wait a minute! I make $4 an hour splitting firewood and all I have to do is pay my mom back for my iPod." I then asked them to write down Washington state's minimum wage: $7.93 per hour.

J.D. King

## Graphing Wages

My students' task the next day was to graph four linear relationships on the same coordinate grid and write equations for each. Again, a linear relationship is one in which change is steady, and it graphs as a straight line. They would graph a day's and then a month's full-time wages for four service industry occupations.

I chose service sector occupations to reflect our community's job market. We do have skilled jobs available in our community. For example, local sheetmetal workers can apprentice at more than $20 an hour with benefits, but the work isn't always steady. Like many places, part-time work is common and the low-wage service sector is growing, as corporations outsource higher-paying manufacturing and skilled jobs to Mexico or China. During the 1970s and '80s, many of the locally owned manufacturing operations were sold to multinational corporations. Subsequently, many of them have moved on. In the last five years alone, the Miller Brewing Co. shut down the 106-year-old Tumwater brewery leaving nearly 400 workers out of a job. Oregon-based Tree-Source closed shop at one of the town's last sawmills and laid off more than 100 workers.

I gave students cards (online at www. rethinkingschools.org/math) with these service sector job titles:

Retail Clerk at Walmart

Security Guard

Retail Clerk at Costco

Home Nursing Aide

I wrote up role cards as though my students had just landed these jobs. For each job, I wanted to paint a picture that showed the importance of and the value of the work. I worded the cards carefully because I knew that many students would have family members making meager livings in these occupations. I didn't want in any way to contribute to the devaluing of human beings and their daily labor.

The Costco card reads: "In high school you went out for basketball. You've always been really strong and light on your feet. In school, teachers asked you to run errands because they knew you'd do the job quickly and well. In your job interview you talked about how you've always loved working as part of a team. It worked. You were hired. Your duties take you all over the store, nearly at a run. You check prices for customers at the register, you return unwanted items, you break down boxes, and you restock with the pneumatic pallet jack."

Students worked in groups of four, with each occupation represented. I told them to read the cards aloud and try to estimate hourly wages for each, given the state and national minimum wages we had discussed the previous day. The task brought up questions. Cody, in typical stream-of-questioning style, asked: "What is a minimum wage? Who gets to decide? What does it mean? I don't get it." Questions like his gave

me an opportunity to talk about the meaning of the minimum wage and share the story of the labor activism that led to Washington state's indexed-to-inflation minimum wage.

Back in 1986, the minimum wage here was $2.30 an hour. The Washington State Labor Council, in collaboration with churches and women's groups, began advocating for an increase. By 1993, after three successive legislative victories, it had risen to $4.90—still too low for a single person to live independently, let alone support a family. Rather than continue to fight for each successive increase, the groups joined forces to lobby to have the minimum wage indexed to inflation. That way future increases would come annually, without expensive and time-consuming lobbying efforts. In 1998, two-thirds of the state's voters passed a measure that set the minimum for the following year at $6.50 and guaranteed a yearly increase, indexed to inflation.

As students talked and asked questions, I learned the word on the street in Tumwater is that Costco is the place to work. Several students had parents who worked as security guards, and many reported relatives in low-paying healthcare roles.

"My mom works at Walmart and she's always working," Stephanie exclaimed. "We never have enough money."

Their discussions assured me that putting algebra into this context would connect with students' lives outside school.

After small group discussions, I passed out approximate hourly wages to go with each occupation. For graphing ease, I rounded wages to the nearest dollar. I set the Walmart wage at $7.00 an hour, which is lower than the company-reported average national wage of $8.23. In many states, however, Walmart starts workers as low as $6.25. While setting it lower than our state's minimum risked confusing students, I did so to expose them to the idea that two different employers such as Costco and Walmart can have different policies that profoundly affect the quality of workers' lives. Reports from the U.S. Department of Labor placed the security guard and nursing aide at $11

# THE DEPENDENT VARIABLE

Following an example that I had prepared in advance, students worked together to draw an x- and y-axis on 11- by 17-inch graph paper. I told them that they would be graphing one day's wages, or eight hours, and that the size of one's paycheck depends on the number of hours worked; therefore, money—the dependent variable—belongs on the y-axis. I said, "I'm not going to tell you what the scale of that axis will be. That depends on your wages, and you'll have to discuss together how high it needs to go. The independent variable, however, will be the same for everyone: You are all going to work an eight-hour day."

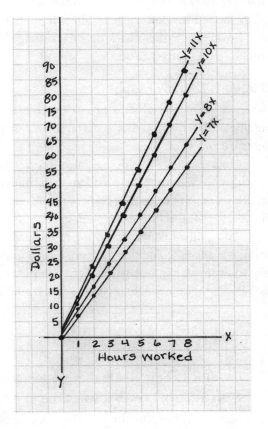

Having students graph all four different wages on the same axes forced them to see that higher wages meant steeper lines. It also helped them see how quickly an extra dollar per hour adds up. After students completed the graphs, I introduced the variables x and y. In this case, y represents your paycheck, and x stands for the number of hours you work. I challenged them to write an equation for each line that would show the relationship between time and earnings. Every group arrived at equations to represent the lines on their graphs.

Later, they would repeat the exercise by graphing a month's pay at the same wage. —J.D.

and $8 respectively. My grocery store cashier provided the source for Costco: $10.

After students presented their graphs, I brought their attention to the algebra involved by asking them to respond in writing to the following prompt: How does the rate of pay affect the shape and steepness of the lines on your coordinate grid? Describe the shape of a graph for the wage of a job at $20 per hour. Describe the shape of a graph for the wage of a job at the federal minimum of $5.15 per hour.

The prompt led students to observe that the steeper the line, the higher the wage, and that each of the situations produced a straight line. Both observations paved the way for introducing "slope"—the rate of increase—and "linear"—a relationship that graphs as a straight line. The prompt also served to identify the coefficient of x or the number that multiplies x as the value that determines the steepness of the line.

Students expressed dismay at the federal minimum wage of $5.15 per hour. The nearly

$40-a-day gap between Costco's $10 per hour and the federal minimum looked enormous. However, they didn't yet have any inkling of how much money it takes to maintain a household; later I would help them back up their outrage by providing that information.

### 'It Sounds Like My Family'

Next, we spent several days practicing recognizing linear patterns in tables and graphs and writing equations from them.

Once students could recognize linear relationships, it was time to broaden their understanding of the fairness of a given wage. A recent film shown on the PBS documentary series *POV* titled "Waging a Living" served my purpose. The online teacher's guide for the program comes with downloadable footage of Jerry, who struggles to get by as a San Francisco security guard.

I discovered later that many students identified with Jerry. Jade wrote: "I remember sitting and listening to Jerry's story and thinking, 'My goodness! It sounds like my family.'"

After hearing Jerry talk about the expense of dressing to work in the fancy lobby of the building he guards, I introduced a problem so my students could experience the costs that come with employment. I proposed that my "lucky" students had just landed jobs. The bad news was that each job required a uniform. To allow students to compare equations, and to emphasize that not everyone's circumstances are the same, I wrote two different uniform descriptions for each occupation. For example, Security Guard A drew the following:

> The company wants you in uniform. Even though you already have nice

## THE Y-INTERCEPT

When I felt comfortable that students were able to recognize linear equations that began at zero, I knew it was time to introduce the concept of y-intercept. This time I had students start by making a table showing pay minus expenses for the first 10 hours of work. This gave students a real-world context for operating with negative numbers.

After they finished the tables, I introduced the standard form for linear equations: $y = mx + b$, in which m represents the pay per hour, and b stands for the expenses. In general terms, m corresponds to the relationship between x and y, and b corresponds the value of y when x equals 0. I asked students to graph pay per hour less expenses and then to write an equation that would describe earnings minus expenses at any given hour. —J.D.

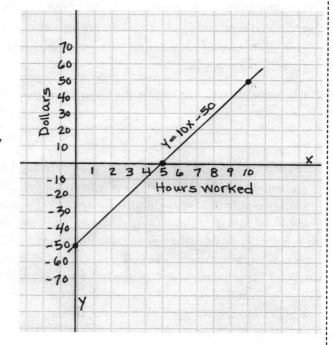

clothes, they issue you the following, at your expense: hat $15, belt $20, pants $36, and three shirts—$25 each.

Security Guard B met with different circumstances:

You're expected to look nice. Luckily you're an average size and you can make it to Goodwill before your first day on the job. You pick up two jackets for $12 each, slacks for $7 and three dress shirts for $5 each. Shoes you have to buy new. You have foot trouble, so you decide you'd better get good ones. They cost you $70.

I assigned occupations to pairs of students and then had each of them draw a different card.

When Angie announced, "Dang! I've been working all day, and I still haven't broken even!" I knew that students were beginning to realize, through the math, that working for a living meant more than paying your mom back for your iPod. They were ready for a bigger problem: rent.

I showed more footage of Jerry in a segment where he described the challenge of living in a long-term occupancy hotel in order to be close to work. Affordable housing outside the city would make him late to work. This time, I had students work in groups of four. I took a page of the classified ads for local rentals and challenged them to find affordable housing.

Heated discussions ensued about whether or not it was fair for a coed group to select the "women-only, no smoking, no drinking" house to share at $250 per month. After they made their housing choices, I introduced additional factors, again on cards. Some were positive, others negative. One card read: "Lucky you! When you moved your grandma's old couch into the apartment, you found $25 in coins beneath the cushions. You start the month $25 ahead." Then I had students use the minimum wage to write an equation that would tell them how many hours they would have to work to make the rent.

For most groups, the number of hours to make the rent ranged from 40 to 80. One group decided that they'd go for the house at $1,500 per month. Given their relatively high Costco wage and the requirement that they pay a month's rent in advance, their calculations revealed that they'd have to work 300 hours to make the rent.

So far, I hadn't complicated our calculations with expenses such as Social Security and employee insurance contributions that shrink a paycheck before it's even cut. I hoped to get to that later. I wanted students to have a solid understanding of how pre-deduction rates of pay are linear. I also wanted to bring students back to their families' and each other's experiences before exposing them to additional expenses.

I framed a discussion around two questions: What happens in families when there's not enough money; and what can happen in a family that makes it so there's not enough money? Again, talk was lively. Students shared family experiences of itinerant homelessness and living with various friends and relatives; struggles to pay the bills; absent, hardworking parents; families split not by rancor but by economic necessity, job loss, death, and disability.

The next day I provided some typical expenses such as $230 for new tires, a $1,500 trip to the dentist for a cracked tooth, $30 for new shoes for a growing adolescent, and $405 each month to feed a family of three. I also discussed the common experience of not getting enough hours of work in a week. Walmart workers, for instance, average 30 hours a week. When I asked students if they knew anyone who has a job but complains about not getting enough hours, most hands went up.

I reminded students of Washington state's minimum wage of $7.93 an hour. We then entered the equation $y = 7.93x$ to our graphing calculators. We looked at both the graphs and tables generated from the equation to answer such questions as, "How many hours do you have to work at the minimum wage to pay for the dentist? For new tires? For new shoes? For food?"

Then I provided students with a print-out of the living wage information for our community. Because the size of household determines expenses, I asked students to use the graphs and tables in their calculators to determine the number of hours at minimum wage it would take to support families of various sizes. They discovered that three people have to work full time at the minimum wage to support a family of two adults and one child. When I asked if this was possible, several students besides Stephanie reported that they never see either parent because they work all the time.

I then asked them to use the mathematical evidence to attack or defend the statement: "Washington state's minimum wage is high enough." Four fifths of my students determined that it wasn't high enough. Mark wrote: "The minimum wage is not high enough. You have to pay the rent, then buy food and what if you have to go to the doctor? How are you going to have enough money to kill some bills? On the minimum wage, you're going to run out of money two weeks before your paycheck and then the rent will be due." Although Mark didn't put math in his writing, his graph had a big black arrow pointing to the gap between income and expenses. On the other hand, Chance thought that the minimum wage was plenty high. He knew money would be tight, but suggested credit cards could help with expenses. He carefully calculated the estimated shortfall during a month requiring new tires: $253. "Just don't have kids until you have a job good enough to pay for it," he advised.

## Impact

Several months after our study of wages I asked students about the impact of the unit. Some students reported that they began planning their futures, including working on good grades and staying in school. Other students wrote about increased awareness of socioeconomic class. "What was interesting was using mathematics to see how other people are in the same status as me," Evan

wrote. "My mom works for just above the minimum and we never have money for extra things."

Lizzie reflected on the teaching and learning as a whole. "It was interesting to learn algebra that way because we learn more than just one thing," she wrote. "It unshields us from the safety of our home to be ready for the outside world."

Next time, I'd like to teach students even more about the world. While I turned my students' attention to the circumstances of people in their own community, those circumstances exist within a larger system permeated by assumptions about the value of labor and the role of the rich and the poor, both locally and globally. I will prompt students to consider factors that make some work worth more than other work and ask them to consider the fairness or unfairness of those factors. I will also challenge them to graph the $1 to $2 a day that most of the world's poor earn and have them consider the hourly wage of the top American CEOs who, according to a *Washington Post* report, bring in up to $40 million per year.

While it's a victory to have students recognize that schooling can improve their personal prospects, next time we'll examine the role that union organizing and labor struggles have played in advancing the right of workers to a living wage. For example, according to an EPI report, on average, workers covered by a union contract have 14.7 percent higher wages, are 28.2 percent more likely to have health coverage, and are 53.9 percent more likely to have pension benefits than nonunion workers.

In their reflections, some students considered their own personal circumstances and others considered the world around them. Their responses let me know that I had met my goals of building a bridge between algebra and the world of wages and work, and of showing them that math can be a tool for reading their world.

My students have learned that algebra matters—and so do they. □

# ACTIVITY BOX

## 'I CAN'T SURVIVE ON $8.25'—USING MATH TO INVESTIGATE MINIMUM WAGE, CEO PAY, AND MORE

BY ERIC (RICO) GUTSTEIN

Sometimes, a slice of reality is so rich that you can teach it at almost any grade level. That's the case with the "McDonald's $8.25 Man and $8.75 Million CEO Shows Pay Gap," recently published on Bloomberg.com [http://www.bloomberg.com/news/2012-12-12/mcdonald-s-8-25-man-and-8-75-million-ceo-shows-pay-gap.html]. The title contrasts the pay of the McDonald's CEO to that of Tyree Johnson, a veteran McDonald's worker making the Illinois minimum wage of $8.25 an hour, but the article tells much more. It also tells us that from 2007, the start of the most recent U.S. economic crisis, until 2011, profits for McDonald's, Walmart, and Yum! Brands (KFC, Pizza Hut, and Taco Bell) soared from 22 percent (Walmart) to 135 percent (McD's). During the same time, the United States lost 1.3 percent of its jobs, but jobs in the fast-food industry increased 7.3 percent—while many older workers who lost good-paying jobs were forced into the fast-food sector, replacing teenagers. These numbers (and more in the article) give plenty of opportunities for students to learn math while using it to more fully understand what's happening around them.

| Company | Profit growth, 2007-2011 | | Compensation for highest paid executive last fiscal year | |
|---|---|---|---|---|
| McDonald's Corp. | ■■■■■■■■ | 135% | ■■■ | $8.8m |
| Yum! brands Inc. | ■■■ | 45 | ■■■■■ | 20.4 |
| Walmart Stores Inc. | ■ | 22 | ■■■■■ | 18.1 |

Source: Data compiled by Bloomberg, company filings, McDonald's Corp.

Math teachers and their students across the grades have many questions to investigate here. In 1st or 2nd grades, students can add dollars and cents and learn multi-column addition. For example, the question—If Mr. Johnson works three hours, how much does he make?—can be solved by adding three 8s (dollars) and three 25s (cents). This can expand to multiplication (If he works 12 hours, much does he make?) as children get older. The numbers lead directly into studying fractions and decimals in upper elementary grades, eventually to multiplying fractions and decimals (37 ½ hours at $8.25/hour) and onto percentages as well (If Mr. Johnson "brings home" only 78 percent of his gross wages due to taxes and Social Security, what is his net pay?).

And, of course, there is the question of the *8.75* million, which primary grade children cannot easily access, but older ones can. Even understanding the numbers can be challenging. The numbers in the article's title look similar: *8.25* and *8.75*. Are they? Are they both decimals? How so? Near each other? In what ways? What is *8.75 million*? How big is it? (see "Understanding Large Numbers," page 110). Can you write it in other ways? Why is it *$8.75 million* in the article's title, but *$8.8 million* in the graph above? Is there a difference? How big?

An obvious question (which the article addresses), is how long does Mr. Johnson need to work to earn what the McDonald's CEO makes in a year (not including stock options, bonuses, and dividends!)? How many of those *8.25s* will fit into that *8.75 million*? If Mr. Johnson works 40 hours a week (which, as the article tells us, few McDonald's workers actually do), and 50 weeks a year, how long to make *8.75 million*? Middle school students can work on scientific notation (*0.825 x 10¹* and *0.875 x 10⁷*), which can lead into division with exponents as well. And they can also confront issues with percents such as "What is the meaning (mathematically, politically, and morally) in the above graph of the *135 percent* profit increase for McDonald's from 2007 to 2011?" and "What do the numbers *7.3 percent* and *-1.3 percent* mean in the graph below?"

What about workers' standard of living over time? The federal minimum wage in 1968 (*$1.60/hour*) is much less than the 2011 level (*$7.25/hour*), so haven't things improved? A naive response might be "yes." But not so fast. The article mentions inflation, and that's an opening for high school students to learn about compound growth, exponentiation, and fractional roots (rational exponents). Inflation calculators (e.g., http://data.bls.gov/cgi-bin/cpicalc.pl) tell us that *$1.60* in 1968 had the equivalent purchasing power of *$10.34* in

| Occupation | Median earning, 2009-2011 | | Employment growth from '07-09 survey to '09-11 | |
|---|---|---|---|---|
| Fast-food worker | ▮▮▮▮▮ | $18,564 | ▮▮▮▮▮▮▮ | 7.3% |
| Child caregiver | ▮▮▮▮▮ | 19,099 | ▮ | 0.1 |
| Cashier | ▮▮▮▮▮ | 20,101 | ▮▮▮ | 2.6 |
| U. S. average | ▮▮▮▮▮▮▮ | 42,110 | ▮ | -1.3 |

Source: Census Bureau's American Community Surveys 2007-2009 and 2009-2011

2011. But what does that mean? How could one calculate it? If you assume inflation as a constant, what is the average rate of inflation over that time period, and what is the mathematics behind this? (*~4.44 percent* annually.)

The article also examines changes in the ratio of "average S&P 500" CEO to minimum wage pay over time, which went from 42 in 1980 to 325 in 2010 and 380 in 2011. High schoolers can examine the below graph and learn linear and quadratic regression to understand the sociopolitical context of CEO to average pay ratios, while predicting future trends.

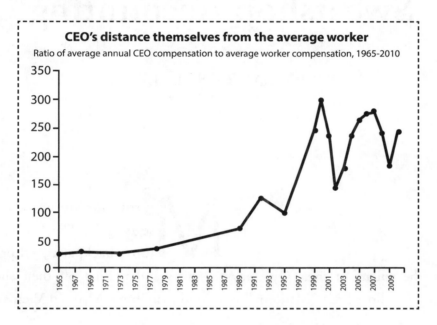

**CEO's distance themselves from the average worker**
Ratio of average annual CEO compensation to average worker compensation, 1965-2010

The article also includes workers' resistance. Mr. Johnson recently joined in the growing national fight for $15 an hour and to unionize fast-food workers. As he says, "I'm trying to fight for what I believe in." After learning mathematics to more fully understand his and others' reality, it would be hard for students not to want to join in and support him.

# CHAPTER TEN

# Sweatshop Accounting

## BY LARRY STEELE

My students are out of their seats again. They're digging into backpacks, flipping down the collars of their Old Navy T-shirts and pulling off their shoes, looking for those little "Made In" tags. Made in China. Made in Vietnam.

"Hey, I was made in Vietnam," says Allan.

The students stick yellow Post-it notes on a big map of the world, marking the countries where their stuff was made. In five minutes, the Post-its obscure the coast of China, plus the Philippines, Malaysia, Indonesia, and other low-wage countries where people struggle to live on as little as $1 per day.

This isn't social studies class. It's business education. The students are examining the connections between their habits as consumers and working conditions in the countries where the goods are made.

As I teach students who hope to be future business leaders, these connections, which are often ignored in business education classes, are the central focus of our classroom work.

Today's high school students can take classes their parents never considered until college, classes like marketing, economics, and international finance. But this 21st-century curriculum needs to be taught as more than a set of technical skills. In our $31 trillion global economy, business decisions affect billions of people. That's why I think it's more important than ever for business education to connect economic skills with the values of democracy, social justice, and environmental sustainability.

That's what my accounting students are doing as they analyze the pattern of yellow Post-it notes on our classroom map. I ask them what they think the pattern shows. "Use what you've learned about accounting," I say.

Allan waves his hand. "Companies are cutting their salary expenses," he says. "They can do it by hiring people in developing countries who will work for not much money."

Allan sees the connection between the numbers on the income statement he is studying and people on the other side of the planet. That's because I supplement our textbook and its columns of worksheet numbers with readings, role plays, videos, discussions, and writing about social responsibility. As we learn how the numbers work, we also take time to look for the people behind the numbers.

"When we format a company's income statement, where do we report net profit?" I ask the class.

After some near-miss answers, Mei Li says, "It goes on the bottom line."

"Right. And what is the bottom line?" I ask her.

"It's total sales minus total expenses," she answers.

"Spoken like an accountant," I say. "But I'm looking for a broader meaning. What does the phrase mean in general. Like, 'The bottom line is you need a college education to get a middle-class job?'"

More near misses, then Mei Li ventures, "The most important thing?"

Bingo.

I want students to think critically about what can happen when profitability becomes the most important thing. Decisions based on business considerations often are presented as if they are "value-neutral"—just part of doing business. Nothing could be further from the truth. I use a lesson called "Sweatshop Math" (page 256), to illustrate the human consequences when business managers focus too narrowly on the bottom line.

To help visualize people living and working in other countries, we watch *Sweating for a T-Shirt,* a video by Medea Benjamin and Global

### A Familiar Example

My students are learning what our accounting textbook calls "the language of business." As they gain fluency, they start talking about debits and credits, business transactions, and financial reports. It's a cross-cultural language spoken by business people worldwide. The key word in this language is "profit."

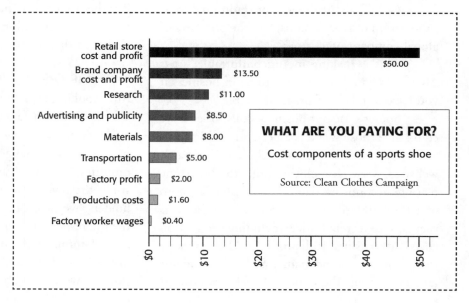

**WHAT ARE YOU PAYING FOR?**

Cost components of a sports shoe

Source: Clean Clothes Campaign

| | |
|---|---|
| Retail store cost and profit | $50.00 |
| Brand company cost and profit | $13.50 |
| Research | $11.00 |
| Advertising and publicity | $8.50 |
| Materials | $8.00 |
| Transportation | $5.00 |
| Factory profit | $2.00 |
| Production costs | $1.60 |
| Factory worker wages | $0.40 |

Exchange (see Resources, page 261). Then, the students read short vignettes from *Rethinking Globalization* about the lives of workers in low-wage countries. Using information from their reading, the students label maps of the world with sum-

## We get a distorted picture when we equate success with higher production.

maries of local working conditions, wage rates, and hours worked. For example, one vignette tells about workers in El Salvador who "get paid 29 cents for each $140 Nike NBA shirt they sew. The drinking water at the factory is contaminated. Women raise their babies on coffee and lemonade because they can't afford milk."

As the students cut and paste labels for their maps, I help them reflect about what they know. Their ideas usually echo views they've heard before.

"If somebody takes a job it's their own choice. It must be better than what they were doing before," says Rob.

"They should go on strike," says Brian.

We talk about the loss of traditional jobs, suppression of labor unions, and the pressures of global competition.

"This is making me uncomfortable," says Alicia. "Maybe I don't want to be in business if you have to take advantage of people."

"So, let's use our business skills to see if we could give those workers a better deal," I say.

We visit the Clean Clothes Campaign website, www.cleanclothes.org, to get some real-life accounting data. The site has a photo of a $100 athletic shoe labeled with all the costs that go into its retail price, including production costs, advertising, and profit for the company that owns the brand name. Soon the students are transferring the data to Excel spreadsheets

and using the program to make colorful bar charts (see an example on page 79).

Brian is shocked. On his chart the 40-cent bar representing factory worker wages is barely visible. There's a long $13.50 bar for the brand-name company, and an even longer $50.00 bar that goes to the retail store. With such a visible and familiar example the students immediately see options for more equitable distribution of income.

"They could pay the factory workers twice as much and it would barely dent the shoe company's share," says Brian.

"Or cut advertising expense by not paying Michael Jordan so much to wear the Nike logo," adds Linh.

"Reduce the retail store's costs," says Desiree. "That's the biggest cost."

I remind Desiree that with her knowledge of fashion, she might want to work in retail sales. "Your paycheck would be part of the retail store's costs. How would you feel about taking a cut in salary?"

"Oh no! They won't be taking it out of MY pocket," she says.

Desiree gets a laugh from her classmates. But the students are beginning to see that global economic connections involve decisions that challenge their own values.

### Mom and Pop vs. the Big Box Store

After looking at how financial decision-making affects people in faraway countries, I want my students to see how doing business "by the numbers" also changes lives in our own communities. As my advanced accounting students study the ownership rights of corporate shareholders, we take time to compare and contrast the interests of shareholders with the interests of local communities. In a unit I call "Mom and Pop vs. the Big Box," we read about competition between neighborhood "mom and pop" stores and the world's biggest "big box" store, Walmart.

We read parts of a recent Walmart annual report. It says the corporation's job "is to see how little we can charge for a product." It describes exciting career opportunities for its workforce of diverse, respectful, well-trained "associates," including 300,000 outside the United States.

Quang is impressed when he finds a list of Walmart financial contributions to Boys' and Girls' Clubs, the United Way, and other local charities in neighborhoods near Walmart stores. I ask him to create a "local citizenship" ratio to compare the amount Walmart spends on community relations with its total sales for a typical store. Quang is less impressed when his calculator shows a ratio that's a tiny fraction of 1 percent.

Next we compare the company's self-portrait to one of the many critical reports about Walmart's business practices around the world. As an example, here is a quote from the website of Walmart Watch, www.walmartwatch.com, an organization that encourages consumers to shop at stores offering employees "good jobs with living wages:"

> There are two ways to cut costs. You can reduce waste and inefficiency. That's great. It's what makes the market system go 'round. But you can also cut costs by putting them off onto someone or something else. … You can muscle down your suppliers' prices, so they have to move production to poor communities and pay wages that won't support a decent life. You can hire part-time workers with no benefits and give them no training. Our taxes and insurance fees will pay for their health care. … You can pressure towns for tax breaks and free roads and water lines and sewers. The other taxpayers will pick up the bill. You can pay only a fraction of the real costs of materials and energy. Nature will eat the damage. This kind of cost-cutting not only imposes injustices on others, it also undermines the market economy. It distorts prices so consumers cannot make rational decisions. It rewards bigness and power, rather than real efficiency.

> "It is a lie that our wages are adequate. Our children don't even have toys. We make baseballs out of old socks. We cannot feed our children right."
>
> **— A garment worker in Nicaragua,**
> National Labor Committee

This statement and Walmart's annual report provide a great opportunity to "compare and contrast" texts. The Walmart Watch text describes how Walmart creates low prices by avoiding costs for labor, health care, and the environment. The costs don't go away, but the company leaves them out of its own accounting. It makes them "external costs," or "externalities," that will be paid later, or by someone else. Walmart avoids taking responsibility for externalities. Our textbook doesn't mention this concept. It might be considered too advanced for a beginning course. I think it's pretty basic, so I introduce it here and return to the concept again and again.

The next day I hand out role-play instructions for a make-believe community meeting, including short character sketches I put together. In this role play, students play Walmart executives, city planning officials, shop owners, shoppers, neighborhood residents, and potential store employees.

When we're ready to begin, I shout, "Action!"

City Planner Mei Li introduces the key speaker.

Quang stands up to make his presentation to the community. He unrolls a large

# HOW DO YOU LIVE...

## ... ON 31 CENTS PER HOUR?

You are a worker at a garment factory in the Las Mercedes Free Trade Zone in Managua, Nicaragua, sewing clothing for major U.S. labels. Your wages, including overtime, incentives, and bonuses, are about 31 cents an hour, or $14.88 per week.

| | |
|---|---|
| Round trip bus fare to work | $2.45 |
| Breakfast (coffee or juice only) | $1.92 |
| Lunch (rice, beans, some cheese, soft drink) | $6.37 |
| Rent (cost for sharing a one-room hut with another family) | $3.53 |
| Water (from one faucet that doesn't always work) | $1.41 |
| **Subtotal** (exceeds your $14.88 in weekly wages): | **$15.04** |
| Electricity | $1.77 |
| Wood (for cooking on outdoor stove) | $1.88 |
| Powdered milk (enough for two infants) | $4.08 |
| Childcare | $6.12 |
| School (textbooks and supplies) | $0.47 |
| **Total:** | **$29.34** |

Now the total is $29.34, without even including clothing, medical care, or food for dinner. The minimum wage in Nicaragua is about 25 cents an hour. How will you survive?

Source: National Labor Committee

sheet of paper representing the plan for a 100,000-square-foot Walmart Super Store.

"We want to build this great store across the street from the Martin Luther King Mall. It will sell everything you need, at low, low prices!"

My students shop at Martin Luther King Mall. They visit hair-styling boutiques owned by African Americans, a dollar store owned by a Vietnamese family, and a dozen tiny restaurants and shops.

"Hey, wait a minute," says Alicia. "My friend's aunt owns a deli sandwich shop on that street. What if all her customers start going to Walmart? She'll go out of business."

"Forget those raggedy stores at King Mall," says bargain-shopper Desiree. "I want to go to one store with good prices. It's convenient."

"Yeah, and they might hire minimum-wage teenagers," says Rob.

Many of the shoppers and residents agree. Some say they've shopped at the local stores all their lives, but even they are impressed by Quang's promise of low prices for everything they need.

Later, reflecting on what we've learned during the role play, the students realize that most of their opinions were based on personal considerations. Nobody stood up and cited Walmart's income statement to show the connection between low prices and low wages. Nobody argued consciously about "externalities," although some students were getting at it when they mentioned shopkeepers that might go out of business if the Super Store opened. I resolve that next time, I'll find a more concrete way for students to correlate their concerns with the financial considerations affecting their community. Maybe I'll cast someone as a labor activist.

Nevertheless, the make-believe community meeting reflected real life. It was clear that by focusing narrowly on a low-cost business model, discount stores like Walmart—and discount shoppers—would sacrifice important social values.

## "Full-Cost" Accounting

What if the accounting system changed to include, rather than exclude, social and environmental costs? From the European Union to Japan to the U.S. Environmental Protection Agency, government agencies and innovative businesses are experimenting with ways to do this. They call it "environmental accounting" or "full-cost accounting."

I want my students to get a taste of how full-cost accounting could work, so I announce that it is snack time. We share slices of crisp, juicy apples from two paper plates labeled "local organic" and "imported." Then, I ask the students to discuss their ideas about the costs to bring each type of apple to the table.

We chart the entire life cycle of the apples, listing all the various costs involved in growing, transporting, and marketing the products. Dividing into small groups, some focusing on organic apples and some on imported varieties, the students compete to see who can make the longest, most complete list of costs. In a second column, the students mark an "I" for costs they think are included in the price of the apple and an "E" for costs they think are externalized.

I hear debates about water, organic fertilizer, and pesticides.

"Where do imported apples come from?" I ask.

"Australia!" "Mexico!" "Chile!" they answer.

"Then don't forget the fuel for long-distance transportation, and the exhaust and traffic jams," I remind them.

We consolidate the group lists on the board. The organic and imported lists look pretty similar—land, water, farm worker labor, a store. But the students begin to find differences, too.

"You can use a truck to get Washington apples to the store, but you need a ship for apples from Australia," says Ramiro.

"Organic fertilizer doesn't pollute, but nonorganic ones might pollute," says Grace.

The groups think most of the costs, even highways and seaports, are included in the price of an apple.

"Diesel fuel taxes pay for highways. The store includes that in the price of the apple," says Rob.

But we find some externalities, too. Business doesn't pay for more air pollution when they burn more fuel. They don't pay if chemical pesticides get into groundwater. If companies were held fully accountable for these costs, they would have stronger reasons to adopt better practices.

Apples provide an accessible example. Full-cost accounting gets far more complicated in industries like forestry, mining, or energy. I ask for some general conclusions and trigger a short debate.

"Companies will have to pay for the damage they do instead of leaving their mess for somebody else to clean up," says one student.

"Yeah, but that will make everything more expensive," argues another.

We try to reconcile these two views and realize that the costs wouldn't change; they'd just be assigned to specific products. That would make the prices more realistic.

"If my business had to pay for all the social and environmental costs, I'd try to treat people and the planet better," says Rushawn.

## Driven by the Market

Do corporate executives really want to promote social justice and environmental sustainability? Maybe there is something in the rules that stops them. The Transnational Capital Auction activity (page 248) reveals part of the answer. Student teams role-play as representatives of developing countries trying to attract foreign investment. They must bid against each other to win new factories and jobs.

As business students, my class knows what investors want: lower costs and higher profits. They quickly learn how to play the game.

"Build your factories in our country!" they say. "Our people are happy to work for $1 per day. We don't allow unions. We'll cut your taxes. Forget about environmental regulations!" Within reason, country-teams that compromise social and environmental conditions to reduce business expenses win the prize of foreign investment. After several rounds of bidding, virtually all of the countries discover that they have been competing in a "race to the bottom"—a race that destroys social and environmental values rather than promoting them.

When the game is over, we talk about why countries enter the race at all.

"Everybody needs jobs," says Irene.

"If they don't have money for their own factories, they've got to get it somewhere," says Henry.

Those are the traditional answers. But our essential question was, "Is there something in the rules that forces a competition to make things worse?"

I explain the concept of "fiduciary responsibility": Corporate executives, bankers, investment advisors, and many kinds of business people are responsible for managing the money invested by other people. Generally, this means increasing the "return on investment" by increasing profits. It means people who are rich enough

# THE GLOBAL CAPITALIST ECONOMY

## CARTOON TEACHING SUGGESTIONS

Show students this cartoon along with other Polyp cartoons (pages 12 and 229). Using data in this cartoon, have students make other cartoons and graphs to demonstrate the maldistribution of income in the world.

Have students do activities from *Rethinking Globalization* (see Resources, page 266) to explore the causes of this unequal distribution of wealth. But these are old data, from the 1980s. Have students research updated figures (e.g., from http://inequality.org/) and update their cartoons and graphs. And the cartoon mixes up wealth and income. Have students investigate and compare data on global inequality of wealth and income, and U.S. inequality as well (see activity on World Wealth Distribution, page 89).

CELEBRATE THE TRIUMPH OF THE **GLOBAL CAPITALIST ECONOMY** WITH THESE FINELY CRAFTED 'INCOME DISTRIBUTION™' SOUVENIR CHAMPAGNE GLASSES

RICHEST FIFTH — 82.7% OF THE WORLD'S WEALTH

11.7%

2.3%

1.9%

POOREST FIFTH — 1.4% OF THE WORLD'S WEALTH

EACH HORIZONTAL BAND REPRESENTS AN EQUAL FIFTH OF THE GLOBAL POPULATION ARRANGED IN ORDER OF INCOME

to have disposable income to invest get richer. People who aren't get their salaries squeezed.

"It's the bottom line again," says Mei Li.

"That's why we won the game," says Rob, whose country now allows child labor and pays starvation wages. "We cut their costs so they could make bigger profits."

There are many reasons for poverty and ecological decline. As we reflect about what we've learned from the Transnational Capital Auction, we talk about all the different ways people struggle to improve their lives. Political parties struggle for democracy against the temptation of corruption. Workers fight to organize unions. Teachers prepare students for more complex jobs.

But as business students, we are studying the financial rules that guide the entire game. I want the students to think critically about whether those rules are creating outcomes we truly desire.

## National Accounting

The U.S. government uses a system of national accounting to measure the performance of our entire country. To understand this, I ask my students to visit the website of the U.S. Bureau of Economic Analysis, www.bea.gov. They download a table that shows the historical size of U.S. gross domestic product (GDP), the dollar value of all goods and services produced in our country. Using graph paper or Excel spreadsheets, they create bar charts that show how the GDP has skyrocketed during our lifetimes to $11 trillion per year. Government leaders, economics textbooks, media commentators, and many teachers agree that the growth of the GDP means our lives are getting better and often equate "growth" with "progress."

A lesson called "What's Up with GDP?" at the website of Facing the Future, www.facingthefuture.org, illustrates the distorted picture we get by equating success with higher production. Students role-play as citizens of a make-believe town called Salmon Bay, Alaska.

Most folks in Salmon Bay earn their living by fishing, but the town also has bankers, lawyers, business owners, retail workers, and the CEO of an oil company. Henry, the banker, passes out everyone's monthly salaries in colored-paper $100 bills. We tape the bills together into chains—shorter ones for the fishing people and longer ones for the professionals and CEO. We use the total income of everyone in Salmon Bay to represent the town's GDP.

Then disaster strikes. A truck hits an oil pipeline running through Salmon Bay, the oil spills, and a thin film flows over the waters of the bay. The fish die. The fishers lose their jobs. People get sick as oil contaminates the water supply.

But the spill boosts business for some companies.

Desiree reads an "after spill" card from the curriculum guide that tells what happens, for example, to retail business owners:

> We are sorry to say that some oil workers and fishers are out of work and are now spending less at the grocery store, movie theaters, and gas stations. However, the good news is that hotels and restaurants have been very busy since the spill, as there are many officials in town reviewing and monitoring the cleanup operations. We are experiencing a 50 percent increase in business. Doctors are busy. Lawyers are busy. The oil company CEO spends millions on the cleanup, and gets a bonus for her performance.

After the oil spill, Henry adjusts pay envelopes for everyone in town. We again measure Salmon Bay's GDP, represented by totaling everyone's colored-paper $100 bills. Allan announces the bottom line:

"The GDP INCREASED!"

This isn't what the students expected. Because traditional GDP measures only economic transactions in dollars and cents, leaving out human suffering and environmental damage, the economy of Salmon Bay looks better after the oil spill.

As always, we remind ourselves of the essential question for this class: What can we see if we look at the lives of people behind the financial numbers? We would see the wrong picture if we used only dollars and cents to account for the success of life in Salmon Bay.

In the future I plan to use more lessons from Facing the Future. They have published a useful student reading guide titled *Global Issues & Sustainable Solutions* that supports the People and the Planet curriculum. This set would help balance any business-education curriculum.

## Redefining Progress

Like full-cost accounting for companies, some economists support full-cost accounting for the national economy. If we want to account for the things that truly make our lives better, what important things could we measure?

After reading about labor struggles, watching videos about environmentally sustainable development, and role-playing as economic decision-makers, my students now can list dozens of important values: "job security," "clean water," "fair pay," "health care," "safe food," "a sustainable future." I scramble to list them all on the board.

The genuine progress indicator (GPI) is an alternative to the GDP index. See www.utahpop. org/gpi.html for a sample document put out by the Utah Population and Environment Coalition. The students discover that it balances many of the important values they listed with more traditional economic indicators. For example, the GPI includes the value of unpaid childcare and housework. It subtracts, rather than adds, the expense of cleaning up pollution.

We print out a chart showing historical growth of the GPI and lay it alongside our earlier charts of historical GDP. The traditional measure of U.S. economic success is increasing faster and faster. But when pollution, crime, natural resource exploitation, and other social and environmental factors are included, the GPI shows that our lives are improving much more slowly.

## The View from the Corner Office

My students are out of their seats again. They're looking out the 40th-floor windows of a "Big Four" accounting firm, the local office of one of the world's four predominant accounting companies. From this height they can see container ships from China unloading at the Port of Seattle. They can see the homeless shelter down the street from the King County Jail. They can see the snowcapped Olympic Mountains 35 miles away across Puget Sound.

The young associate accountants who meet us are plugged-in and enthusiastic. Their clients are medium-sized businesses and large corporations. It's a people-oriented profession, they say. Everyone has a laptop computer with wireless connections to their colleagues, plus links to accounting rules and tax laws in every country of the world. They love the variety, pay, and perks of their jobs. The best news is that Big Four firms often recruit business-school students during their senior year of college. Study accounting, pass the CPA exam, and you could start your career earning $30,000 to 40,000, with great upward potential, according to the American Institute of Certified Public Accountants.

Some of my students will find a place in corporate accounting. Some may start their own accounting practices and help small businesses or individuals. Many will use their knowledge of accounting principles to be informed investors and consumers—and perhaps activists.

Whatever paths they choose, I hope they have the courage to ask the critical questions that lurk behind every financial decision: When we measure success in terms of dollars and cents, who do our decisions affect, and how? □

---

All students' names have been changed.

# GLOBALIZATION, LABOR, AND THE ENVIRONMENT

## A LINEAR PROGRAMMING UNIT

BY INDIGO ESMONDE AND JESSICA QUINDEL

Story problems often ask students to put themselves in someone else's shoes and to make decisions for them. Students are discouraged from questioning the contexts of these problems and from using their values and outside knowledge to make sense of them.

We decided to design a six-to-eight-week linear programming unit (based on the "Cookies" lesson in the Integrated Mathematics Program of the National Council of Teachers of Mathematics) that encourages students to challenge the problem context and come up with more equitable contexts of their own.

The following problem is one of the unit problems, which students work on for several weeks. Their job is to advise the CEO of a shoe manufacturing company.

sWOOsh Inc. is a new shoe manufacturer trying to introduce their product into the marketplace. They have decided to do an initial manufacturing run of two types of shoes: a basketball shoe and a skateboarding shoe. They need to decide how many of each type of shoe to make.

### OPTION 1: INDONESIA

They pay workers about $750 to make 100 pairs of basketball shoes in this nonunionized factory in Indonesia and $375 to make 100 pairs of skateboarding shoes. (The workers are paid by the hour, so these are estimates based on how long it takes to make each kind of shoe.)

To make 100 pairs of basketball shoes, they need 50 square feet of synthetic material and 50 square feet of leather. To make 100 pairs of skateboarding shoes, they need 100 square feet of synthetic material.

The duties and shipping for one flat of 100 pairs of basketball shoes is the same as the cost for one flat of 100 pairs of skateboarding shoes: $700.

sWOOsh is sure that they can sell as many shoes as they produce. Their decision is limited by the following constraints: They have budgeted $15,000 for labor and $17,500 for shipping; they have 2,200 square feet of synthetic material, and 900 square feet of leather.

They want to make the most profit possible. The manufacturer's profit is $1,200 for 100 pairs of basketball shoes and $800 for 100 pairs of skateboarding shoes. How many flats of each type of shoe should they make?

As students learn about linear inequalities, we want them to also learn about global inequalities. Therefore we provide texts and videos for students to learn about globalization and the economic and environmental impact of sweatshop manufacturing.

### OPTION 2: CALIFORNIA

The shoes are manufactured at a union shop in California: Labor costs increase, shipping/duties decrease, materials are the same, and profits decrease. Students start to understand how the choice of different amounts for the variables influences the outcomes for the situation.

### EXPLORING FURTHER

Above is just one example of a problem that students explore in this unit. Other problems that students explore help them to understand the mathematics of linear programming and to learn more about economics and globalization. We provide text and video resources for students to do background research into these topics. For the culminating project in this unit, we organize a debate in which students make recommendations about where the shoes should be produced and under what conditions. Students represent environmentalists, domestic labor, Indonesian labor groups, and corporate interests and are required to use mathematical evidence in their arguments. ■

# POVERTY AND WORLD WEALTH

**RECOGNIZING INEQUALITY**

BY SUSAN HERSH AND BOB PETERSON

An important part of a person's understanding of global issues is the recognition of the dramatic inequalities between nations and social classes within countries. Math is an essential tool for acquiring this understanding.

The purpose of this activity is to demonstrate graphically the vast differences in wealth between different areas of the world. It combines math, geography, writing, and social studies.

We remind students of some of the things we learned about colonialism, such as how great quantities of silver and gold were stolen from the Americas and taken to Europe. We also explain that current relations between countries and international organizations, such as the World Trade Organization, also affect how much wealth countries possess. We make sure that students know the following terms: resources, GNP, wealth, distribution, income, power, and colonialism. (Additional teaching ideas that help set the context for this lesson can be found in *Rethinking Globalization.* See Resources, page 266.)

## MATERIALS

• 11" x 17" world maps for each student or pair of students.

• 50 chips (25 of one color and 25 of another) for each map.

• 25 slips of paper with "I was born in [name of continent, based on chart]."

• 25 chocolate chip cookies.

• World map laid out on playground, or signs with names of continents and yarn to distinguish boundaries.

• Transparency of "World Population and Wealth" table (page 92).

• Six "negotiator" signs with yarn to hang around students' necks.

• Writing paper and pens or pencils.

• Additional cookies for students who don't get any during the simulation (optional).

• Worksheets (optional) for students to write down their estimates (available at www.rethinkingschools.org/math).

## SUGGESTED PROCEDURE

1. Give each student or pair of students a world map. Have them identify the continents and other places you have been studying.

2. Ask students how many people they think are in the world. After students have guessed, show them an almanac or a website with a current estimate. Ask: If we represent all the people in the world with 25 chips, how many people is each chip worth? (For 7 billion people, for example, each chip would represent 280 million people.)

3. Give 25 chips to each student/group and have them stack them on the continents, based on where they think people live. Have students write down their estimates using the worksheets mentioned in the materials list or on a piece of paper. Discuss student estimates and then tell them the accurate figures. Have them rearrange their chips to reflect the facts. Ask students what the differing stacks of chips tell them about the world's population.

4.    Explain that you are now going to give them another 25 chips of a different color and that they represent all the wealth produced in the world (the monetary worth of all the goods and services produced every year, from health care to automobiles). Each chip therefore represents 1/25 of the world's total amount of goods and services produced. Tell the students to put the chips on the continents to indicate their estimate of who gets this wealth.

## Teachers can connect students' feelings about fairness to the data on world wealth.

5.    Discuss student estimates and record them on the chalkboard. Have students reflect on the sizes of the two different sets of chip stacks, representing population and resources. Collect the chips.

6.    Tell students you are going to demonstrate how population and wealth are distributed by continent. Have each student pick an "I was born in …" slip from a container labeled "chance of birth." Students may not trade slips. (As you distribute the slips, listen for stereotypical reactions to the continents—these will be useful in the follow-up discussion and will indicate possibilities for future lessons.)

7.    Have students go to an area that you have designated to represent that continent. (Playground maps work great for this.) After students are in their areas, remind them that they each represent about 280 million people and that you are going to distribute the world's wealth.

Have each continent/group designate one person to be a "traveling negotiator" and distribute a traveling negotiator sign to those people.

8.    Explain that once the bag of resources is passed out to a representative from each continent, each group needs to sit in a circle and discuss their situation. Tell the students there will be a cross-continent negotiation session, then a time for the traveling negotiators to return to their home base to discuss their negotiations with the rest of their group, and finally a time for any trading or donating of resources. Students on each continent are to talk about how many resources they have compared to people of other continents and to discuss ways they might negotiate to increase their resources. They may plead and/or promise. (Note: Every continent, except North America, will have at least one "stay-at-home negotiator" and one traveling negotiator. The North American person can stay put or travel throughout the world. Also note that because Mexico is by most definitions part of North America, I explain to the students that for the purpose of this simulation we will be using "Latin America," which includes South America, Central America, the Caribbean, and Mexico; instead of "North America," we will use "United States/Canada.")

9.    Use a popular treat that can easily be divided in half—such as chocolate chip cookies—and distribute them according to the percentages noted in the chart. Announce the number of treats you are giving to each continent as you do so. Provide a paper bag for each continent to keep their treats in. As you dramatically place each of the resources into the bag, remind students they are

not to eat the treats until after the negotiation session.

10. Announce that the negotiation session is to begin. Only traveling negotiators may move to a different continent. When they come, they should sit in a circle with the stay-at-home negotiators and discuss the distribution of wealth and what should be done about it.

11. After five or 10 minutes, tell all traveling negotiators to return to their home continents. Each group should then discuss the negotiations. After a few minutes, announce that the trading session may begin, and if a continent wishes to trade or donate resources, they may. After that, instruct the people holding the resource bags to distribute the resources to people in their group.

12. Give each continental group tag board and markers. Tell them to make some signs that describe what they think of the way the resources were distributed.

13. Bring students back together for a whole-class discussion. Have each group share their posters and perspectives. Show students the information from the world wealth chart via a transparency or handout. Connect their emotions and feelings of fairness to the information on the chart. (At this time, a teacher might give out additional treats to those students who did not get any.)

### QUESTIONS WORTH POSING
### IF THE STUDENTS DON'T ASK THEM
### THEMSELVES

• How did the distribution of wealth get to be so unequal?

• What does the inequality of wealth mean in terms of the kinds of lives people lead?

• Who do you think decides how wealth is distributed?

• Should wealth be distributed equally?

• Do you think that, within a particular continent or nation, wealth is distributed fairly?

• How does the unequal distribution of wealth affect the power that groups of people hold?

• Within our community, is wealth distributed fairly?

• What can be done about the unequal way wealth is distributed?

• Who can we talk with to find out more information about these matters?

If your students have studied colonialism, ask them what role they think colonialism played in creating this inequality.

After the discussion, have students write an essay about their feelings, what they learned, what questions they continue to have, and what they might want to do about world poverty. Some students might also make wall posters that graphically depict the inequality of wealth.

### FOLLOW-UP ACTIVITIES

A few days after this simulation, "Ten Chairs of Inequality" (page 213) is a useful activity to help students understand that wealth is also unequally distributed in individual countries.

Students also can do follow-up research on related topics, such as: the role colonialism played in the wealth disparity; how current policies of U.S. corporations and the U.S. government affect people in poorer nations; the role of groups such as the WTO and the International Monetary Fund; and what different organizations and politicians are doing about world poverty. (Refer to the list of "Organizations and Websites for Global Justice" in *Rethinking Globalization*. See Resources, page 266.)

Three notes of caution with this activity: First, as with any simulation (or role play) this should be understood to be just that—a simulation. We can in no way reenact the violence of poverty and hunger that kills tens of thousands of children daily. We are providing a mere glimpse. Second, while Africa and other areas south of the equator do not have lots of wealth as defined by GNP, those areas have great human and natural resources and this fact should not be lost on the students. Finally, in this simulation we seek to describe rather than to explain current power and wealth arrangements. They can, however, be powerful tools in motivating students to want to figure out the answer to the essential question: Why? ■

# WORLD POPULATION AND WEALTH

## DATA TABLE

| CONTINENT | POPULATION (in millions) | % OF WORLD POPULATION | # OF STUDENTS | | WEALTH (GNP) (in billions of dollars) | % OF WORLD GNP | # OF TREATS | |
|---|---|---|---|---|---|---|---|---|
| | | | (in a class of 25) | (in a class of 30) | | | (out of 25) | (out of 30) |
| Africa | 1,022 | 14.8 | 4 | 4 | 1,701.5 | 2.6 | 0.5 | 1 |
| Asia | 4,164 | 60.4 | 15 | 18 | 20,737.7 | 31.6 | 8 | 9.5 |
| Oceania | 36.5 | 0.5 | 0 | 0 | 1,255.5 | 1.9 | 0.5 | 0.5 |
| Europe | 738 | 10.7 | 3 | 3 | 20,119.9 | 30.6 | 7.5 | 9 |
| U.S. and Canada | 344.5 | 5 | 1 | 2 | 16,719 | 25.5 | 6.5 | 8 |
| Latin America | 590 | 8.6 | 2 | 3 | 5,101.4 | 7.8 | 2 | 2 |
| WORLD TOTAL | 6,895 | 100 | 25 | 30 | 65,635 | 100 | 25 | 30 |

Sources: World population figures are from the United Nations, Department of Economic and Social Affairs, Population Division, http://esa.un.org/unpd/wpp/Excel-Data/population.htm based on figures from The 2010 Revision of the World Population Prospects. GNP figures are from the World Bank, http://data.worldbank.org/indicator/NY.GNP.ATLS.CD, for 2011. For countries where 2011 figures were not available, figures from the next available previous year were used.

Gross National Product (GNP) is defined as the total national output of goods and services produced by a nation and its citizens in a particular year. Percentage of world wealth is an estimate based on total GNP.

For purposes of this chart, divisions of population and GNP by regions followed the population divisions for the United Nations. Europe includes the Russian Federation and Greenland. Asia includes the Middle East. Latin America includes Mexico, the Caribbean Islands, and South America. Oceania includes Australia and New Zealand.

Please note: We have included in the above table, calculations for a classroom of 25 students and of 30 students. If one uses 30 students, the lesson needs to be adjusted accordingly.

# UNEQUAL DISTRIBUTION OF WEALTH IN THE UNITED STATES

## RECOGNIZING INEQUALITY, PART TWO

BY MICHAEL LANGYEL

Display the graphic "U.S. Households and Wealth" (page 94). Explain to students that there is a difference between income and wealth. Economist Edward Wolff, author of *Top Heavy: A Study of the Increasing Inequality of Wealth in America* argues that to analyze economic inequalities, one must go beyond the annual income of a person or family and look at their wealth, which is defined as the dollar value of assets, minus any debts or liabilities, held by a household at any one time. (Income, by comparison, refers to the flow of dollars over a period of time, usually a year.)

Ask the students what the house graphics represent and what the money-bag graphics represent. Ask: "What do these data show us?"

Tell students that they are to represent this data in three-dimensional form using 100 pennies (or chips) and a chart of 100 squares.

Have the students draw out a 10 x 10 chart with squares large enough that a penny can fit entirely inside a single square. (For a downloadable PDF of a chart like this, see www.rethinkingschools.org/math.)

Tell the students that each square represents one percent of the households in the United States, and each penny represents one percent of the country's wealth.

Have the students work individually or in groups of two or three to construct their three-dimensional graph—stacking pennies or chips on squares to represent the distribution of wealth among different segments of the U.S. population.

### ADDITIONAL PROCEDURE

For students needing more explicit instructions, suggest the following:

1. First divide the 10 x 10 grid in three regions by coloring each with different colors:

   • 1 square = richest 1%
   • 19 squares = the next richest 19%
   • 80 squares: the remaining 80% of the families.

2. Next divide the pennies into three piles: 43 pennies for the richest square, 50 pennies for the next region, and 7 pennies for the third region.

Note: Some students may have difficulty stacking 43 pennies on the richest square. Also some will struggle to divide 7 pennies among the 80 squares in the largest region.

**QUESTIONS**

Have students reflect in writing or in discussion on these questions:

1. What do these data tell us about the distribution of wealth in the United States?

2. Why do you think wealth is distributed this way?

3. What is the relationship between wealth and poverty?

4. What government policies on spending and taxes do you think the three different groups might support or oppose? What other data might be useful to more deeply understand the distribution of wealth in the United States? ∎

# UNEQUAL DISTRIBUTION OF WEALTH

## U.S. HOUSEHOLDS AND WEALTH

1% of U.S. households own **43**% of the wealth

19% of U.S. households own **50**% of the wealth

80% of U.S. households own **7**% of the wealth

Source: *Recent Trends in Household Wealth in the United States: Rising Debt and the Middle-Class Squeeze—an Update to 2007, by Edward N. Wolff. Working Paper No. 589, March 2010. Levy Economics Institute of Bard College, www.levyinstitute.org.*

# CHAPTER ELEVEN

# Chicanos Have Math in Their Blood

## Pre-Columbian Mathematics

BY LUIS ORTIZ-FRANCO

Mathematics education dates its beginnings to the time when human beings began to quantify the objects and phenomena in their lives. Although the process of counting (one, two, three ...) was the same for different groups of people around the world, the symbols by which they represented specific quantities varied according to their own particular cultural conventions. Thus, the Babylonians, Romans, Hindus, Egyptians, Angolans, Chinese, Aztecs, Incas, Mayas, and other groups each wrote numbers differently.

Likewise, cultures that achieved a level of mathematical sophistication that allowed them to manipulate their number symbols to add, subtract, multiply, and divide and to perform other algorithms did so in different ways. Today, even within a single society, various groups of people (for example, accountants, physicists, engineers, mathematicians, chemists, and so on) view and manipulate mathematical quantities differently from one another. The study of the particular way that specific cultural or ethno groups—whether they are different national, ethnic, linguistic, age, or occupational groups or subgroups—go about the tasks of classifying, ordering, counting, measuring, and otherwise mathematizing their environment is called ethnomathematics.

The ethnomathematics of pre-Columbian cultures is a topic frequently overlooked in discussions about the cultural achievements of pre-Columbian civilizations, and omitted from college-level textbooks on the history of mathematics. It is important that we focus on pre-Columbian mathematics: Such a focus broadens our perspectives on pre-Columbian cultures and may stimulate us to integrate new perspec-

## Chicanos' impressive history of cultural achievement has been almost entirely ignored in U.S. schools.

tives and topics into our classroom teaching. The ultimate beneficiaries of these educational practices will be North American society in general and North American children in particular.

The integration into school mathematics of pre-Columbian mathematics is important for both political and mathematical reasons. The teaching of the mathematical traditions of pre-Columbian cultures can contribute to achieving a crucial political goal: infusing multiculturalism into education. Students will thereby develop an appreciation for the diverse ways different cultures understand and perform mathematical tasks. This will expose students to the sophisticated mathematical traditions of other cultures and demonstrate that performing mathematics is a universal human activity.

For Chicano students in particular, studying pre-Columbian mathematics will allow them to learn more about their ancestors in both cultural and mathematical contexts. This integrated approach can do much to instill pride in their culture and also increase their confidence in their ability to learn and do mathematics and, perhaps, later participate in mathematics-based careers.

Chicanos are mestizo, a blend of European and Mexican Indian ancestry. In this country, the cultural roots of Chicanos in pre-Columbian cultures are acknowledged at the social level but are usually ignored in the American educational system. In a scene in the movie *Stand and Deliver*, teacher Jaime Escalante attempts to motivate his Chicano mathematics students at Garfield High School in East Los Angeles by saying: "You burros have math in your blood." Escalante's comment surely will seem sadly ironic to readers familiar with the statistics regarding the low educational achievement of many Chicano students, and familiar with both the history of mathematics in Maya and Aztec societies and the relatively minuscule number of Chicanos who pursue mathematics-based careers in the United States.

Despite a long and distinguished heritage in the sciences, arts, and letters in their own culture, Chicanos are one of the least educated groups in this country. Their impressive history of cultural achievement has been almost entirely ignored in U.S. schools for at least two reasons. First, the Western orientation of the educational process largely disregards the achievements of conquered indigenous civilizations and their descendants, such as Mexican people. Second, as a result of this ethnocentric orientation, many teachers and other school officials in the United States are unaware of the mathematical accomplishments of pre-Columbian societies.

This article discusses a pre-Columbian number system that played an important role in the cultural activities—such as commerce and dating historical events—and the development of Mesoamerica. (The term "Mesoamerica" refers to the geographical region that encompasses the area from northern-central Mexico to northern Costa Rica.) In this essay, I discuss the Mesoamerican number system and the origins of the vigesimal number system, and make some instructional suggestions.

## The Mesoamerican Number System

The Mesoamerican number system is a positional vigesimal (that is, base 20) system. It employs only three symbols to write any whole number from zero to whatever quantity is desired. This symbol represents 0:

The dot represents the quantity of 1:  •
And the horizontal bar represents the quantity of 5: ▬

To write the numbers 0 to 19 in this system, the two processes of grouping and addition are used. Numbers 2 to 4 are written using an addition process. For the number 5, five dots are grouped into a horizontal bar. Numbers 6 to 19 are written using the addition process (see Figure 1).

For numbers larger than 19, a vertical positional convention is used (see Figure 2). In this convention, the bottom level is for the units, the next level up is for the 20s, the third level up is for the 400s (1 x 20 x 20), the fourth level up is for the 8,000s (1 x 20 x 20 x 20), and so on in powers of 20. For instance to write 20, we write the symbol for 0 in the first level and a dot in the second level. To write 65, we write three dots in the second level (3 x 20 = 60) and a horizontal bar, representing five units, in the first level.

This vigesimal system of numeration is practical to use and can easily be adapted to classroom instruction. For example, the operations of addition and subtraction are relatively straightforward processes. In the case of addition, one has to remember that since 20 units in a lower level are equivalent to one unit in the next level up, 20 units in a lower level are replaced with one dot in the level above it. Figure 3 shows the sum of 8,095 plus 1,166, before and after the grouping process.

This addition example points to rich mathematical experiences in which students can be engaged. Unfortunately, beyond addition, subtraction, and associated algorithms, we do not know whether the Mesoamerican civilizations knew how to multiply, divide, or perform other mathematical algorithms with this vigesimal system. But we do know that the Mayas wrote books on paper just as we do. We know that for 1,600 years before Columbus accidentally arrived in the "New World," the Mayas wrote and kept thousands of books in which they recorded their history and cultural achievements.

Tragically, however, the Spanish conquerors and missionaries burned and otherwise destroyed all of the Mayan libraries and archives. Possibly some of those destroyed books contained information on algorithms and other mathematical systems that pre-Columbian societies devised. We know that Mayan astronomers had calculated the cycles of the heavens so exactly that they could predict solar and lunar eclipses to the day, hundreds of years in advance. For example, a Mayan astronomer predicted, some 1,200 years in advance, the solar eclipse that occurred on July 11, 1991.

The Mayas knew the synodical revolution of Venus, and some scholars argue that the Mayas also knew the synodical period of Mars and perhaps had parallel knowledge about Mercury, Jupiter, and Saturn. Given their ability to make these calculations as well as to predict celestial phenomena, it is reasonable to believe that they knew how to perform mathematical algorithms other than addition and subtraction.

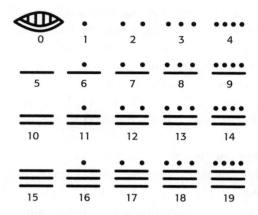

**Figure 1:** MESOAMERICAN NUMBERS 0–19.

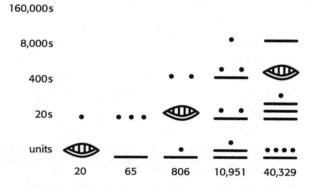

**Figure 2:** EXAMPLES OF MESOAMERICAN NUMBERS BEYOND 19.

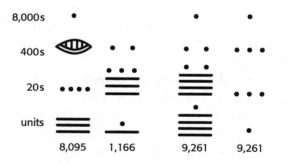

**Figure 3:** THE SUM OF 8,095 PLUS 1,166, SHOWING THE RESULT BEFORE AND AFTER REGROUPING.

This belief is rooted in the origins and uses of their vigesimal number systems.

## Origins of the Vigesimal Number System

Archaeologists and other scholars maintain that humans first inhabited North America around 30,000 years ago and, in particular, Mexico about 9,000 years later. Groups of hunters and gatherers roamed Mesoamerica for thousands of years before they became sedentary. Jacques Soustelle pins down the advent of agriculture in Mesoamerica at approximately 4000 B.C. However, the organized life that can be called civilization in the region began approximately 5,000 years ago. The social evolution of Mesoamerica can be traced from the hunter-gatherers through the successive civilizations of the Olmecs, Zapotecs, Mayas, Toltecs, Aztecs, and so forth.

The earliest evidence of numerical inscriptions that used positional systems of bars and dots has been traced to the Olmecs in approximately 1200 B.C. This date is significant, since some 800 years before Aristotle, Plato, and Euclid (whose society did not have a positional number system) began making contributions to Western culture, the Olmecs were already using a positional system. It is worth noting that it was not until 499 A.D. that the Hindu-Arabic number notation using zero in a positional convention first occurred.

The Zapotecs of Oaxaca used the Mesoamerican vigesimal system in their calendars between 900 and 400 B.C. (Between 400 B.C. and 300 B.C., the Izapan culture used the same convention. Later, the Mayas, to whom the vigesimal system is mistakenly attributed, used this system extensively, between 199 A.D. and 900 A.D. The Mayas developed their amazingly complex calendar system and astronomical sciences around this mathematical system hundreds of years before the achievements of Galileo and Copernicus.

## Recommendations to Teachers

The pre-Columbian positional number system can be taught at various educational levels. As we have already seen, it could be included in the elementary school curriculum as a way to deepen how students understand the decimal system. In fact, the Mesoamerican system may be easier for children to grasp than the decimal system: The vigesimal system is visual, and the representation of quantities involves only three symbols—0, 1, and 5—and manipulative materials can be adapted to give them physical representation.

For example, Dienes blocks can be adapted to the Mesoamerican base 20 system. Teachers can assign to the smallest blocks the value 1, to the intermediate-sized blocks the value 5, and to the larger blocks or to a group of four intermediate-sized blocks the value 20. Alternatively, in classrooms that do not have Dienes blocks but have manipulative materials of different colors and sizes (rods or chips), different colors or sizes of rods or chips can be used for 1, 5, and 20.

At the middle school and high school levels, discussion of this number system can be included in social studies classes as well as mathematics classes, to broaden student appreciation of the cultural achievements of ancient peoples and the fate of conquered civilizations. Teachers can use this topic to illustrate that impressive mathematical achievements of Mesoamerican civilizations were ignored, devalued, or destroyed as part of the rationale for subjugation and domination. For presentations in social studies courses, a map of Mesoamerica is indispensable and can be obtained from the National Geographic Society (for additional sources, see Resources, page 269).

---

# ACTIVITY BOX

## THE HIDDEN GRAIN IN MEAT

BY STEPHANIE KEMPF

One billion of the world's people do not get enough to eat, yet half the grain grown in the world is fed to livestock. Why? To fatten the cattle up for sale to people who can afford to buy meat. Chronically hungry people rarely have the money to buy meat.

Most cattle today do not graze freely on pasture grasses—if they did, their meat would be leaner and healthier. Instead, they are penned up in crowded "feedlots" and given large quantities of grain. The meat from grain-fed cattle is higher in fat.

For every 16 pounds of grain fed to a cow, we get only one pound back in meat on our plates. Producing that pound of meat requires 2,500 gallons of water. In many areas of the world, people do not have access to even a small amount of clean drinking water and must walk miles a day to get it.

### DO THE MATH

If your entire class went to McDonald's and each student ate one Quarter-Pounder, how much grain was used to produce the class's lunch? How much water was used?

Explain why you think this is or is not a problem. If it is a problem, what are possible solutions?

---

From *Finding Solutions to Hunger: Kids Can Make a Difference*. See Resources, page 267.

Teachers can consult other references for more details on the historical origins and uses of this numerical system and for more information on the pre-Columbian cultures who used it.

In mathematics classes at the middle school and high school levels, students can explore interesting mathematics through the Mesoamerican vigesimal system. Teachers can devise exercises comparing the polynomial representation of numbers in our decimal system and the vigesimal system. For example, students can explore interesting mathematics by relating powers of 10 to the value of digits in numerals in the decimal number system and the value of units in the vigesimal system: While the value of a digit in the decimal system is multiplied by a power of 10 that corresponds to the place of the digit of the numeral, the value of the same number of units in the same corresponding place in the vigesimal system is multiplied by a power of 20. This can lead to discussions about powers of 20 as a product of powers of two and powers of 10, to illustrate that the value of units in the vigesimal system increases exponentially faster than the value of the corresponding units in the decimal system. This in turn can serve as a natural introduction to topics related to exponential growth and exponential functions.

Furthermore, in mathematics classes where students are already proficient with the usual algorithm for multiplication in the decimal system (in grades 5 through 12), teachers can also include classroom activities or homework assignments requiring students to use their creativity when working with the vigesimal system. For instance, the teacher could break the class into groups of three or four students each and ask the groups to generate ideas of how to carry out multiplication in this number system. This idea can be extended to include division as well. These challenging assignments may turn into group projects that can last for an extended period of time.

Given the sophisticated system of ancient Mesoamerican mathematics and the gross underrepresentation of Chicanos in mathematics-based careers in the United States, the comment of Jaime Escalante to his students is indeed sardonic. The legacy of racist discrimination against the cultures and native peoples of Mesoamerica, which resulted from the military conquests and colonization ushered in by Columbus's arrival in the "New World," has continued to this day in the imperialistic practices of U.S. society. It has resulted in an educational system in this country which effectively ignores the rich tradition of excellence in mathematics in the Chicano students' background and fails to instill in young Chicanos a sense of pride in their heritage and a positive self-image.

Despite all this, some modern Chicano mathematicians have made valuable contributions to applied and abstract mathematics—David Sánchez, Richard Griego, Manuel Berriozabal, Richard Tapia, and Bill Velez, to name a few. Their contributions should be used to encourage Chicano students to pursue the exceptional mathematical heritage of their pre-Columbian ancestors. ☐

---

This article was adapted from a version which first appeared in *Radical Teacher* magazine.

# CHAPTER TWELVE

# Whose Community Is This?

## Understanding the Mathematics of Neighborhood Displacement

### BY ERIC (RICO) GUTSTEIN

The equation went up on the board as my 12th-grade "math for social justice" class silently and soberly stared at it.

$$150,000 - 291,000 = 92,000$$

I talked as I wrote:

Rico: "You've paid $291,000, on a $150,000 mortgage, and you still owe $92,000. Check that math out. That's good math, let's look at that math. One hundred and fifty thousand dollars minus $291,000 equals $92,000. Look at that math. (I pause for 20 seconds as students look, and mumble to themselves and neighbors.) Think about that. Hey! You started with a 150, you paid 291, and you still owe $92,000. Good math, huh? What's going on here?"

Antoine: "They're taking your money."

Daphne: "The bank is taking advantage of you and taking your money."

Rico: "This is legal—this is how banks lend money and make money."

(I pause, and repeat it slowly.) "This is legal—this is how banks lend money and make money."

I asked students, "What are some questions you could ask here?" Renee said, "Why is it legal?" Daphne asked why didn't more people look into it so they wouldn't end up in that situation.

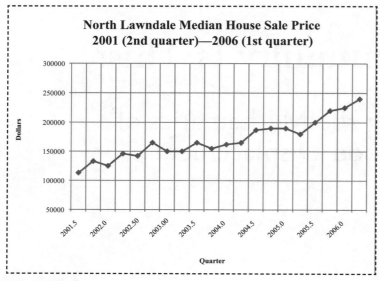

**North Lawndale Median House Sale Price 2001 (2nd quarter)—2006 (1st quarter)**

**Figure 1**

So went a typical day in this class, one in which everything we did during the year was to learn and use mathematics to study students' social reality. We did this for students to understand the root causes of oppression in their lives—to *read the world*—to prepare them to be able to change it— *write the world*—as they see fit.

## Our Setting

We were in the Social Justice High School ("Sojo") in Chicago's Lawndale community. Sojo was born through a multiyear battle in the 1990s for a new high school in Little Village—a densely populated, large, Mexican immigrant community with one overcrowded high school. A victorious 19-day hunger strike by neighborhood activists in 2001 [See David Stovall, "Communities Struggle to Make Small Serve All," *Rethinking Schools*, Summer 2005, Vol. 19, No. 4] forced the district to build the new school, and in 2005, Sojo and three other small high schools opened

in the campus (~375 students each). Though the building is in Little Village (South Lawndale), 30 percent of the students are black, from North Lawndale, a bordering community, and the other 70 percent are Latina/o. Almost all students are low-income (~97 percent), as Lawndale, a spiritually and culturally rich community with deep reservoirs of resilience, is also economically battered. In a climate of extreme education privatization, nationally and especially in Chicago [see Jitu Brown, Rico Gutstein, and Pauline Lipman, "Arne Duncan and the Chicago Success Story: Myth or Reality?" *Rethinking Schools*, Spring 2009, Vol. 23, No. 3], we proudly proclaim that Sojo is a quality *neighborhood public school*, not a charter, alternative, or selective-enrollment school. Any student in Lawndale can attend any of the four schools. I was part of Sojo's design team and have worked there since 2003, and I taught this class in the 2008–09 school year.

## Studying Neighborhood Displacement

Displacement was part of students' realities—gentrification in North Lawndale, deportation in Little Village, and foreclosures in both. I started the unit by telling the story (with family permission) of Carmen, a student in class. Her grandmother paid off her North Lawndale mortgage years before, but because of rising property taxes and a leaking roof, took out a subprime (adjustable) home equity loan. When the rate set upwards, she lost the house. The families of two other students in class were struggling to stay in their houses, and some students discussed their lives and observations, as boarded-up homes were all around. Through these discussions and analyzing house prices (see Figure 1), their questions emerged: Will we be able to stay here? From where and how does gentrification arise? What is the original purpose or plan? Why our neighborhoods? Where are families supposed to go?

To study displacement, students initially learned *discrete dynamical systems* (DDS). A DDS is complex, and even simple ones can behave chaotically. A DDS has at least one baseline and one recursive equation, and one can use them to produce a mortgage or credit card schedule. More complicated DDSs have multiple sets of equations, with which students learned to model HIV-AIDS transmission in Lawndale later that year. For example, the monthly payment on a $150,000, 30-year, fixed-rate mortgage at 6 percent annually is $899.33. Its DDS is:

$a_0 = 150,000.00$ [$a_n$ represents balance due at start of month $n$]

$a_{n+1} = a_n + .005a_n - 899.33$ [what you owe at the start of a month is what you owed the previous month, plus the interest on what you owed, minus your monthly payment.]

My plan, specifically about mortgages as part of the larger unit, was for students to unpack them and see how much more than the actual cost of a house one actually pays over the years. In addition, I wanted students to understand how subprime (high-cost) mortgages worked, as well as the relationships of interest to principal, concepts like negative amortization, and more. My concern was that students begin to appreciate how banking worked as part of a larger capitalist economic system and its relationship to their lives and experiences.

I started by teaching students to use a DDS to model an interest-bearing savings account. The following describes the day that ended with the discussion that starts this article. We began by reviewing homework: create the DDS and find the balance after one year on a $500 deposit at 3 percent annual interest. Marisol wrote on the board:

$a_0 = 500.00$

$a_{n+1} = a_n + .0025 \times a_n$

Using our overhead graphing calculator, she showed the results after one month, then up to 12, and said, "So overall in a year you're gonna be left with $515.21."

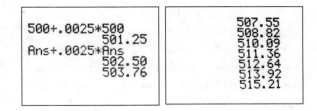

After some back and forth with her clarifying, we moved on. The next problem was: "Assume you have this *$500* deposit that pays *3 percent* per year, but you withdraw *$25* a month. Create the DDS—when will you run out of money?" Vanessa tackled this and said, "OK, I did the same as Marisol, but I subtracted the $25 because you also withdraw 25 a month," and showed us we'd run out of money in 20 months:

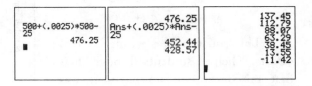

These preparations took us to the reality of students' communities, in which foreclosures had tripled over three years. The week before, students examined the graph in Figure 2 and discussed how mathematics reflected what they were seeing and experiencing.

Next was to find how much a median-income family (in each Lawndale) could afford for housing without "hardship," using HUD's 30 percent guideline. Minerva said that a median-income Little Village family—$32,317—could afford $807.92 monthly, explaining, "Eight hundred and eight dollars. I divide their annual income by 12 to get how much they earn per month and multiply that by .3 to get 30 percent."

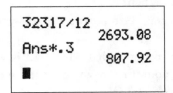

So far, the mathematics hadn't challenged students much. But my next question did:

If a median-income family in *your* community (either Lawndale) takes out a 30-year, fixed-rate $150,000 mortgage at 6 percent interest:

1) Create a DDS with $a_0$ being $150,000 and 6 percent interest annually;
2) Determine whether they can afford the mortgage "without hardship." If yes, how large a mortgage can they afford? If no, explain why not.

As we worked as a whole class, students' initial attempt was:

$$a_0 = 150,000$$
$$a_{n+1} = a_n + .06a_n$$

This equation was wrong on two counts. I helped students through their first misconception: no payment was subtracted. Then Antoine said that the *.06* was annual and should be *.005* (monthly). Daphne, Ann, and Vanessa then discussed the meaning of each term in the equation. I stayed out of it until I pushed Vanessa to explained every symbol that Ann had entered into the overhead calculator:

```
150000+.005*150000
-808
           149942.00
■
```

Vanessa: ... Your monthly payment, and $149,942 is how much you owe.

Rico: When?

Vanessa: After you give—your first payment.

Rico: Exactly, that $149,942 is what you owe after your first payment. OK, how much less than your initial payment is what you owe now?

Antoine: $58.

Rico: $58. How much did you pay? The first payment of $808, $750 goes to interest to the bank. Only $58 reduces your loan balance. Understand how ... banking works. Almost ... 15 times as much as reducing the payment goes to the interest, which is profit for the bank. Yes, they have to pay their employees. That's how they make money.

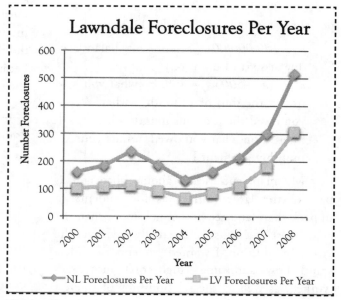

**Figure 2**

I was trying to interweave students' reality, mathematics, and a broader political and economic analysis, but didn't belabor the point and returned to the math, knowing that students would shortly uncover the "strange equation" that begins this article. Ann then taught students how to set up the equation in the calculator so that each time she pressed ENTER, the calculator would show the next month's balance. I interrupted to refocus us:

"So what is the problem? What is the question? What are we trying to do here?"

Daphne: Are we trying to see how long before they pay it off?

Rico: Not only are we try—

Carleton: Can they afford it!

Rico: Yes, we're trying to see, can they afford this mortgage. How many payments are there altogether?

Calvin: 360.

I had Ann press the calculator button 360 times as she counted aloud while we watched the balance shrink on the projector. After 360 presses, the board showed:

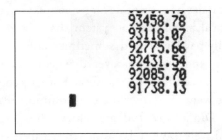

After 30 years, the family still owed almost $92,000. I asked what was the total paid after 30 years and, on the overhead calculator, multiplied 360 by $808, roughly $291,000. That prompted the dialogue and equation that starts this article.

The lesson here: No median-income Little Village (or North Lawndale) family could afford this mortgage without hardship. And, as we had discovered when we examined neighborhood prices, many houses were this much or more—especially new condos. A student brought in an advertisement for a new North Lawndale condo for $285,000, not including the $10,000 gated parking space.

Over the next few days, I had students answer two related questions:

- What income *is* needed to afford a mortgage of $150,000—a 30-year, fixed-rate, 6 percent annual interest mortgage?
- What mortgage amount, with the same terms, can a median-income North Lawndale/Little Village family afford?

By graphing the DDS on their calculators and adjusting their numbers, students uncovered that a $150,000 mortgage would require roughly a $36,000 income and that median-income families in Little Village and North Lawndale could afford mortgages of $134,750 and $84,500, respectively—not $150,000.

Although students saw and lived with displacement, understanding how and why it worked involved examining complex global processes of finance capitalism. The two Lawndales experienced displacement in ways both similar and different, due to real estate development, the economic crash, (un)authorized immigrant status, geography, housing stock, deindustrialization, and proximity to public transportation and highways. As the unit continued, I told students about a plan to build a massive gentrification complex in Little Village. Antoine was very interested, and, upon my suggestion, we investigated after school. Through internet research, we uncovered that a transnational capital investment fund was trolling the globe for investors to profit from displacing him from his community. Antoine was livid and presented his findings to the class along with a four-minute promotional video for the proposed development.

Students also delved into the math of subprime mortgages to better understand how banks profited from them. They were fascinated with this topic. I had students create a scenario, including some negative amortization, for each of three types of subprime mortgages: adjustable-rate, pay-option, and balloon mortgages. They were to turn in, for each:

1. the mortgage terms as they changed—i.e., interest rate, amount borrowed, monthly payment, and number of months;
2. a table showing how much the borrower paid at different time periods, at what interest rate;
3. what the borrower paid over the loan term, itemizing principal, interest, and refinancing costs, if a balloon; and
4. a comparison of the amount from #3 with the cost of a 30-year, fixed-rate, 6 percent per year mortgage.

In the second part of the unit—immigration and deportation—students investigated the U.S. government and NAFTA's role in displacing Mexican farmers from their land, to the *maquiladoras* along the U.S.-Mexico border, and eventually to Little Village. They studied the table, below, and a graph that showed the concentration of USDA subsidies to agribusiness and how, after NAFTA, the price paid for Mexican corn dropped while the cost of tortillas soared. They also analyzed data on Mexican immigration. (See Figures 3 and 4, page 107)

As a way to "write the world," students held two public forums (one in each neighborhood) and shared what they learned with family, friends, and community. The students who presented on immigration showed the following slide and concluded with a causal argument:

**NAFTA's Impacts on Mexico:**

- Corn subsidies to large U.S. corn growers
- Cheaper to produce U.S. corn
- Cheaper to import U.S. corn than grow in Mexico
- Mexican government stops corn price supports
- Mexican government stops tortilla price controls
- Immigration increase to the United States

## Reflections—My Students' and Mine

Developing and teaching a curriculum that supported learning about these interlocking sociopolitical processes *and* college preparatory mathematics challenged me, despite my experience with critical math. My academic goal was for students to learn serious math—precalculus, algebra, discrete mathematics, quantitative reasoning, and statistical analysis—to better understand neighborhood displacement and more, while appreciating math's usefulness in doing so. It's nonnegotiable that students learn math to have full opportunities for education, life, and individual/community survival. Social justice

mathematics demands this, as part of supporting marginalized communities' self-determination, and educators cannot shortchange students' mathematical learning or life chances.

My sociopolitical goal was, as much as possible, to have students understand the causes, mechanisms, and roots of displacement in each neighborhood. I wanted them to see that both Lawndales had the same larger context—a global political and financial system that played itself out in particular and sometimes contradictory ways—and thus whatever differences students saw between their own and each other's communities were far outweighed by commonalities.

Did all my students achieve this? I cannot definitively say, and I want to clarify that not all students learned the same, based on my assessments of their mathematics learning and sociopolitical understandings. In hindsight, creating and teaching an interdisciplinary curriculum through which students could unwind the interconnections of neighborhood displacement was more complicated than I anticipated for both students and me. Nonetheless, students expressed that they learned a lot. In a reflection, Mónica, a Latina and lifelong Little Village resident, wrote:

> Some connections that I see between these two parts of the unit are that in both communities, people are being forced out of their homes. Of course, it's different situations, but similar causes. African Americans are being forced out of their homes because they can't pay for their homes. The taxes go up so much that they can't afford to keep living in those communities, so they are forced to look for another place to live. For Mexican people, the problem is that they don't have jobs in Mexico because corn isn't being sold, because it's cheaper to import subsidized U.S. corn than to grow their own. That forces Mexicans to leave their family and homes to come to the U.S. to look for a job.

... Also, the house mortgages don't only affect one community, but both. They are sometimes the target of bad loans that only make banks richer! I want the people in my community to know that we are really similar in these situations. That there is more that makes us similar, less that makes us different. If we want to fight the bigger people out there, the best way is to unite. Fighting each other is not going to take us anywhere. I think this is something *very* important our community should know.

Renee's comments were particularly powerful in linking the unit to her reality as a Latina:

The unit made many relations between black and brown communities. There are so many misconceptions about black people as well as brown. Both communities think bad about each other. In the black community, they might say that Mexican people don't belong in this country because we're illegal aliens. As well, there are Mexican people who say that all black people have a Link card and spend all their money on clothes, etc. What people don't understand is that we both have the same struggles. They might seem different because of the color of our skin but deep down inside our parents struggle to get by with sicknesses, drug addictions, or unemployment. People are dealing with foreclosures and then become homeless. ... When we did the 30 percent of the median household income for both communities we figured that we can't

**Concentration of USDA Corn Subsidies to Agribusiness**

| Pct. of Recipients | Pct. of Payments | Number of Recipients | Total Payments 1995-2006 | Payment per Recipient |
|---|---|---|---|---|
| Top 1% | 19% | 15,729 | $10,726,604,754 | $681,964 |
| Top 2% | 30% | 31,458 | $17,053,420,149 | $542,101 |
| Top 3% | 39% | 47,187 | $21,870,918,998 | $463,495 |
| Top 4% | 46% | 62,916 | $25,767,405,826 | $409,553 |
| Top 5% | 52% | 78,645 | $29,022,040,929 | $369,026 |
| Top 17% | 84% | 267,395 | $46,941,027,794 | $175,549 |
| Top 18% | 85% | 283,124 | $47,629,179,204 | $168,227 |
| Top 19% | 86% | 298,853 | $48,258,099,906 | $161,478 |
| Top 20% | 87% | 314,582 | $48,834,286,526 | $155,235 |
| Remaining 80% of recips. | 13% | 1,258,332 | $7,336,588,731 | $5,830 |

**Figure 3**

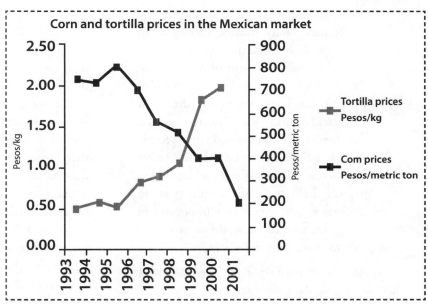

**Figure 4**

afford the house that we are living in. Our family members kill themselves in factories trying to make ends meet. This unit taught us that we have the same struggle. People always ask what similarities do we all have, and this unit tells us why we are the same.

As for how mathematics helped them, Carleton explained what he learned, critiqued the system, and wrote about helping others:

Learning the dynamical system helped me really understand how and why people were losing their homes. It showed how small of an amount of income the average black/brown family was making and how since it was a small amount, how hard it was to pay off the mortgage loans. Not only was it hard but the banks were really stealing money from these people because they would end up paying a double amount of money than they took out because of this thief called Interest. ... These were also the most important things that I learned because they helped me understand what I know now. It helped me to be able to predict whether or not a family would be able to pay off their loan with certain types of mortgages, and this is very important in being able to read the world so we can be able to share with the world.

And Renee used what she learned to analyze her own particular circumstances:

It's crazy how banks give you this loan with a monthly payment that eventually people don't really get out of debt. There are people who don't know this and believe in their capitalist country. People need to know what happens, why they get into debt, especially that what the banks do is legal with our government. People such as my sister lose their homes because they don't read the papers they are signing when they get a loan for a

mortgage. People need to know the difference between the different loans that are out there. The only question that I have after this unit is can what the banks do be made illegal?

The most helpful part of this unit was the dynamical systems. As soon as I really learned how to work with the dynamical systems I came home and grabbed my credit card bill and the mortgage and plugged them in the calculator. Paying the minimum balance on my credit card wasn't enough. I would have to pay double my minimum balance to get out of it in less time. Obviously, what my mother is paying isn't enough to finish paying the house in 30 years. The worst part about this is that what she pays isn't 30 percent of her income, it's more.

Despite these and other students' claims of understanding DDSs and finding them useful to read the world, learning the math, with conceptual understanding, was not easy. Almost all my students had attended under-resourced neighborhood schools that, despite many teachers' efforts, insufficiently prepared most for advanced mathematics. Some students in class had solid mathematical backgrounds, some quite shaky ones, and others in between. However, I believe two factors contributed to students' persistence. First, students were overall engaged in the year's topics, which they chose to study and which were personally meaningful—the 2008 presidential election, HIV/AIDS, criminalization of youth/people of color, and sexism. Second, math provided a way to understand their lives and answer their questions, such as: Whose community is this?

## Taking It to the Community

The student-led public presentations (which they titled "Our People, Our Issues: Math as Our Weapon") were at the end of the year, one in each neighborhood. Students created the 81-slide presentation, with my minimal feedback and editing, and although we didn't have time for a full run-through, and students were anxious the first night, they presented what they had learned and received much positive feedback. In written reflections, students expressed feeling good that they had informed their families and communities, applauded their own and classmates' collective presentations, and critiqued themselves for insufficient preparation, nervousness, and reading too much from their PowerPoint. To a person, students thought it worthwhile and important to teach their people. From my point of view, providing students the opportunity to "take it to the community" and share what they learned is an important component of teaching for social justice and lets students develop competencies that they will need as future agents of change.

### Don't Let Them Pit Us Against Each Other!

The theme that both Lawndales had more commonalities than differences surfaced strongly at the forums. The group that presented on immigration/deportation went last, with a fitting final slide:

### Why Should We Care?

- Both communities face the same problems but different situations.
- There are many lies and stereotypes about both Mexicans and African Americans.
- "Mexicans steal the jobs of U.S. citizens."
- "African Americans are lazy."
- Don't let them pit us against each other!

### Conclusion

It was important that students came to understand, as Renee said, that they had the same struggle, in a context of "divide and conquer." Racism has long impacted Lawndale, and racial division between African Americans and Latinas/os is one of its many effects—in Sojo's first years, black students were sometimes unwelcome in Little Village and even attacked after school. The displacement unit supported black-brown unity (as people here call it) by having students analyze the sociopolitical conditions of their lives—through mathematics—and realize that they have common enemies and a common struggle. As Mónica wrote, the best way is to unite to "fight the bigger people out there," and not each other. Though not all students experienced or knew much about each other's neighborhoods, I am confident that all students left class with a deeper understanding of their commonalities. To re-emphasize: This unit was based on interconnected mathematical and political analyses. In my view, this provides a basis for multiracial solidarity, which we will need in order to read and write the world—with and without mathematics. □

All students' names have been changed.

# CHAPTER THIRTEEN

# Understanding Large Numbers

BY BOB PETERSON

"It's a zillion!" calls out one student in response to my question of how to read the number "1,000,000,000" I've written on the overhead. "No, it's not, it's a million," argues another.

Despite the fact that my 5th graders have been taught place value throughout their elementary years, there is something about big numbers that lends itself to guessing. Perhaps it's the omnipresent state lottery advertisements that tend to blur big numbers together. Or more likely it's the fact that big numbers are just difficult to read, much less understand. Imagining a billion boggles my mind, whether I'm trying to fathom the number of galaxies swirling around the universe or the number of molecules in a drop of water.

Kids are fascinated by big numbers, especially if they connect with the real world. Thanks to the U.S. government's addiction to military spending, students have an endless stream of large numbers to study. And those numbers just get bigger and bigger!

The growth in the military budget comes as schools face massive budget cuts. Teaching about these matters provides students an opportunity to improve their understanding of large numbers and, more importantly, to understand the power of math in debates about the future of our communities and world.

Before I delve into budget issues with my students, I do a couple of activities to help them put meaning behind place value. One year, the night before I was to start my mini-unit on big numbers and budgets, the students' homework was to ask family members what they thought one million and one billion meant. The next day the students shared their responses, which ranged from the precise, "One million is one thousand thousands;" to the comical, "It's what you get when you win the lottery;" to the practical, "It's enough money to buy everything we ever need and still have some left over."

I asked the students, "How many days equals a million seconds?" After some initial guesses, the students worked in groups with calculators using different strategies to solve the problem. Eventually they came up with about 11.6 days.

I then asked how long it would take for us to count to a million. Some students suggested we just do it and time ourselves. Others were more skeptical. After some practice with six-digit numbers we estimated that it would take, on the average, about two seconds a number. Some more calculation and the class realized it would take a little over 23 days. "I'm not going to be wasting my time doing that," one student proclaimed.

To visualize a million, I asked the students to look closely at a strand of their hair and then told them that a million of those piled on top of each other would reach up to a seven-story building. I also showed the students the book *How*

*Much Is a Million?* by David Schwartz. Some of the pages are filled with tiny stars—14,364 per page. The book encourages students to guess how many pages of stars it would take to reach a million, and they are surprised to find it would take 70 pages.

I then repeated some of these activities with a billion. The students soon discovered that their calculators did not go that high, so we did some

## Students can learn to understand the power of math in debates about the future of our world.

whole-class work. We calculated that one billion seconds equals about 32 years. After timing the counting of a series of very large numbers, we estimated it would take about three seconds a number to count to a billion—leading us to conclude it would take almost 96 years to count to a billion. We examined the star-filled pages of *How Much Is a Million?* and calculated that it would take 70,000 such pages!

### Billions for War, Millions for Schools

Next I wrote $80,000,000,000 on the chalk board and had a student read it. I then wrote the number $10,000,000 and did the same. I had students guess the significance of those numbers and then explained that the former is the estimated cost of war and occupation of Iraq for a year, and the latter is approximately the amount of money that is going to be spent renovating and adding to our 100-year-old building. I asked, "How many schools like ours could get a major renovation for the cost of just one year of the Iraqi war?" After a wide range of guesses, I asked how we could know for sure.

The class decided to go around the room counting by 10 million and that Markese would keep track of the number of school construction projects that could be bought. As each student added another 10 million, Markese made a mark on a piece of paper. After we went around the class two times, we had only reached 460 million and it was clear our effort to count to 80 billion was going to take a while. When we finally reached one billion, Markese announced we could rebuild 100 schools. I stopped the counting and suggested the students use mental math to figure out the final solution. "8,000!" one student called out. "That's a lot of schools that could be rebuilt!" added another. "That's more than in all of Milwaukee!" added a third. In fact, a quick check with the State Department of Public Instruction showed that there are just over 2,100 public schools in all of Wisconsin.

Later, as part of this mini-unit, the students discussed graphs from United for a Fair Economy (UFE) that contrast the U.S. military budget with federal social spending and with the military budgets of other countries around the world. Using data found on the internet, students figured out that one Stealth Bomber, at the cost of $2.1 billion, could have paid for the annual salary/benefits of 38,000 teachers. This was of extra significance to my students because our school principal had just informed their

parents we'd be losing two teachers (gym and music), half our librarian time, and two paraprofessionals. "Just for a little part of one of those bombers, we could have all our teachers back," one student said.

I did this unit at the very end of the school year, in the midst of news of mounting budget cuts and continuing conflict in Iraq. I wanted to give my departing 5th graders a different perspective on the cuts: Too often the talk about budgets is filled with hopelessness and inevitability. I knew that in just a few days we'd only touched the surface of a complicated issue. I resolved that during the next school year I would cover this earlier, so that we could spend more time looking at how policymakers make budget decisions, the relative merits of various types of spending, and what social action groups are doing on these issues.

Holding students' attention during the waning days of the school year is always a challenge. This lesson held their attention, and it was a fitting way to end the school year as the students prepared to go to middle schools that also faced drastic cuts.

"This really isn't fair," one student wrote when I asked the class to reflect on the matter in their journals. "So much money is being spent when our schools need so little."

Actually our school—like most schools—needs a lot. It's just that the amount is so little compared to the hundreds of billions of dollars this nation spends each year on the military budget.

Hmmm, I wonder how many schools that could rebuild ... and how many jobs that would create? □

## FOR MORE INFORMATION ON MILITARY SPENDING

### ORGANIZATIONS AND THEIR WEBSITES

**Center for Popular Economics,** www.populareconomics.org

**Center on Budget and Policy Priorities,** www.cbpp.org

**Children's Defense Fund,** www.childrensdefense.org

**Cost of War,** www.costofwar.com

**Dollars and Sense,** www.dollarsandsense.org

**National Priorities Project,** http://nationalpriorities.org/

**United for a Fair Economy,** www.ufenet.org

**War Resisters League,** www.warresisters.org

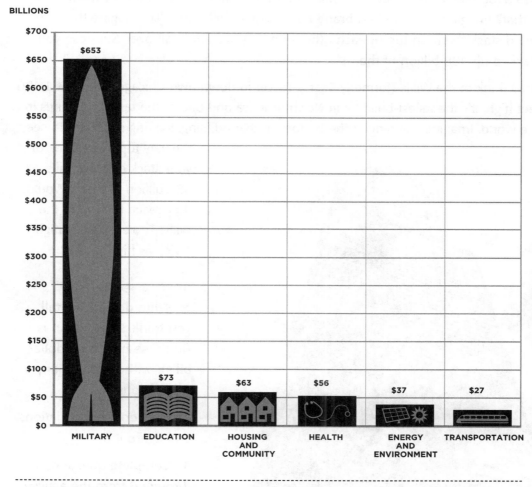

# MILITARY VS. SOCIAL SPENDING

## FISCAL YEAR 2013 DISCRETIONARY BUDGET

BILLIONS

Source: *http://nationalpriorities.org*

# ACTIVITY BOX

## HOW MANY WILLIS (SEARS) TOWERS TALL IS $488 BILLION?

BY ERIC (RICO) GUTSTEIN

The amount of money requested by the Presidents (Obama and Bush) and appropriated by Congress for the war in Afghanistan, through the end of fiscal year 2011 was $488 billion (adjusted for inflation). A lot of money, no? But how much really is that? Imagine that you had brand new crisp $1 bills and also imagine that you could stack them on top of each other and they wouldn't fall over. Suppose you stacked up 488 billion of them!

Now think of the Willis (formerly Sears) Tower in downtown Chicago. At about 1,450 feet high, it's the tallest building in North America and one of the tallest buildings in the world. Imagine standing at the bottom of that building, looking up, trying to see the very top. Now think of your stack of dollar bills, all 488 billion of them. Would they reach the top of the Willis Tower? Half as high? Twice as high? More?

### DO THE MATH

1. Write down how tall you think $488 billion is (in those nice crisp dollar bills), without doing any calculations.

2. Now do the calculations and figure it out.

3. Complete questions 2 through 6 from the activity "How Much Does the War Cost?" (page 40).

*Source: http://costofwar.com*

# CHAPTER 14

# When Equal Isn't Fair

## Using Ratios to Scale Up Mathematical Arguments

BY JULIA M. AGUIRRE
AND MARIA DEL ROSARIO ZAVALA

| | |
|---|---|
| Roberto: | That's not fair! |
| Ms. Julia: | What's not fair? |
| Roberto: | If she pays $1,500, that is half of what she makes in a month. |
| Ms. Julia: | But the sister is paying the same as the others. How is that not fair? |
| Roberto: | Yes, but they do not pay as much. It is not the same for the older sister and the brother. |
| Ms. Julia: | How do you know? |
| Roberto: | I'll show you. |

Roberto, a bright and soft-spoken Latino youth, sparked a conversation that used mathematics to examine issues of fairness. As he showed his classmates his mathematical reasoning and the various ratios and percentages that justified his claim about fairness, it was like wildfire as mathematical ideas traveled across the room.

Roberto and 40 other soon-to-be 9th graders were participating in a special five-week summer mathematics program that emphasized mathematical enrichment and college readiness. We planned and co-taught the math classes with four other mathematics educators in three classrooms for this summer school enrichment program. We attempted to infuse important aspects of teaching for social justice to help students deepen their mathematical understanding and engagement.

Our students were mostly African American, Latina/o, Vietnamese, and East African immigrant youth who enrolled in our program for reasons ranging from a need to be better prepared for high school mathematics, to a parental desire for their children to get more mathematics practice. All were slated to attend high school in an urban school district in the Pacific Northwest.

This lesson was part of a two-week unit on proportional reasoning. Rather than using a commercially available math curriculum, the math team created activities on a weekly basis using a variety of resources that emphasized problem-solving and connections to real-world applications.

During this proportional reasoning unit, many of our students struggled to reach a key conceptual milestone—moving from additive/absolute reasoning to multiplicative/relative reasoning. Additive reasoning emphasizes values as independent of and unrelated to anything else, while multiplicative reasoning requires explicit consideration of how measures or values are related to each other. For example, in the circle problem below (Lamon, 1999), an *additive* answer to the question "Which has more circles?" would be to say that each picture has the same number of circles and so neither has more, while a *multiplicative* answer could be that Picture 2 has more since the ratio of circles to triangles is 1:1, while the ratio of circles to pentagons in Picture 1 is 3:4. We wanted our students to have some experiences with multiplicative reasoning to broaden their mathematical reasoning and because of how important that kind of thinking would be for

algebra, the class that most students would be enrolled in their freshman year.

1. Which picture has more circles?

Picture One

Picture Two

After two days (about six hours of class time) our students still maintained an additive perspective when attempting to solve problems. For example, in the snake problem (adapted from Lamon, 1999) below, the students maintained that the snakes grew the same amount: three feet; rather than comparing the change in growth: one snake growing 3/4 of its height, while the other snake growing 3/5 of its height. They struggled with looking at how quantities were proportioned focusing on how adding or subtracting could get them an answer.

## Pet Snakes

Two years ago, Jo measured her pet snakes String Bean and Slim. String Bean was 4 feet long and Slim was 5 feet long. Today they are both fully grown. At her full length, String Bean is 8 feet long. Slim's full length is 8 feet. Did they grow the same amount? Did one grow more?

As the math team debated how to address our students' firm stance in additive reasoning, Julia recalled a recent family situation when planning her father's 70th birthday party that seemed to embody this mathematical dilemma. Our team developed the Papi's birthday problem (Figure 1) based on this story to make progress on two key goals of teaching mathematics for social justice:

1. developing classical math knowledge.
2. using math as an analytical tool to understand and critique issues of power and inequity.

Our thinking was that the Papi's birthday problem was an accessible story that our kids could relate to because it is ultimately about how to figure out how each person can pay a fair share. We believed that our students would be interested in the income of each person given several students had voiced aspirations of becoming a medical doctor. Furthermore, we believed that some students might be sensitive to single-parent family structures and budgeting (we had several students living with one parent and discussed work and household budget issues in class) as well as relate to big family celebrations because some had shared recent experiences with family reunions and quinceañeras.

Sitting in our planning meeting, it was clear to us, as teachers, that the proposal the brother makes is not fair, and that a fair solution involves reasoning proportionally around party costs per person based on their monthly salary. We agreed in our group that we were not morally apathetic about the mathematical outcome; rather we wanted students to confront their notions of fairness when numbers were "equal." We believed that providing at least two different positions would spark discussion and invite mathematical debate, hopefully leading to using proportional reasoning to justify their solution. We also believed that some students might think

of another way to determine a fair share, so we wrote the task to include other possible financial arrangements. Still, we talked as a group about how to facilitate the task and thought it best to not disclose our position to begin with.

### Initial Student Positions on Fairness

After launching the problem, a few students immediately dismissed both proposals and felt that the brother should pay for everything because he made the most money. Anticipating this solution, we told students that both sisters felt strongly about sharing the cost. With that alternative eliminated, students were forced to consider the proposals offered in the problem.

Many students quickly decided that the brother's idea was most fair because everyone pays the same. Fair and equal meant each sibling should pay the same amount—$1,500. The mathematical evidence was a division equation:

$$4,500 \div 3 = 1,500$$

Students insisted that it would be unfair if each sibling paid different amounts. They initially rejected the university professor's suggestion.

### Pushing Proportional Relationships

After initial discussions about fairness, we encouraged our students to write down their mathematical ideas including equations or other

**FIGURE 1:** THE PAPI'S BIRTHDAY TASK AS PRESENTED TO OUR STUDENTS.

---

### PAPI'S 70TH BIRTHDAY

A true story

It was Señor Aguirre's 70th birthday. His three children wanted to throw him a big party to celebrate. The hall rental, mariachi, food, and decorations will cost a total of $4,500. The brother, a special medical doctor (anesthesiologist) who makes about $20,000 per month, suggested that the three children split the cost equally.

One of the sisters, a university professor who makes about $6,000 per month, said that would not be fair. She suggested the following: the brother pays $3,150. She would pay $900, and the other sister, a partner in the family business and single mom of two boys who makes about $3,000 per month, should pay $450.

---

Write a position statement using mathematical evidence (e.g., proportions, ratios, percent) to support your conclusion to the following questions:

• Which person do you agree with and why?

• What is fair in this situation?

• Can you think of an alternative financial arrangement that might be better (more fair)?

number relationships that might help them take a position. We wanted students to push themselves to see if the situation could be worked out in more ways than one. Initially, the room was mostly quiet as students re-read the problem, or began "crunching numbers," sometimes by subtracting or dividing. We prompted our students to share ideas to push on their proportional reasoning. We asked students to write down their position using mathematical evidence and to use numbers and equations that expressed the relationships as part of their mathematical evidence. We directed them to explicitly state the kind of mathematical evidence we expected to see: "proportions, ratios, percent." However, it was up to the students to represent the mathematics in their solution and justify their position.

The task question, "Which person do you agree with and why?" was one we hoped would prompt students to formulate mathematical justifications. Since most students' first solutions agreed with the brother's proposal, they wrote 4,500 / 3 or 4,500 ÷ 3 = 1,500, or each person pays 33 1/3 percent. We verbally prompted them to explain what 4,500, 3, 1.500 or 33 1/3 percent meant in relation to the problem from a mathematical standpoint and then further probed what this meant in relation to fairness as represented in our second question, "What does

fair mean in this situation?" We asked: "What would it mean if they paid the same amount given the information about income, career, and family in the problem?" We wanted them to look at their solution *in relation to* the contextual information. However, we did not explicitly ask them to revisit their claims of fairness after this second prompt.

Justification is an important mathematical practice. We pushed students to justify their solution including why they did not select other alternatives. By extending the justification to explain why other alternatives were not as fair, students could make stronger mathematical connections about why their solution made sense and was defensible.

In this way we hoped students would create ratios for comparison. In the excerpt above, Roberto stated an important mathematical relationship that challenged his original position of taking the brother's side. Referring to the other sister, he said, "If she pays $1,500, that is half of what she makes in a month." When we asked him to justify, he produced the following ratios:

**Brother's way:**

| | | |
|---|---|---|
| Other Sister | 1,500 / 3,000 | = .50 |
| Professor Sister | 1,500 / 6,000 | = .25 |
| Brother | 1,500 / 20,000 | = .075 |

When further prompted, he labeled each component of the ratio *cost share per monthly income*, and confidently stated that the other sister's cost share would be half of her monthly income. Roberto then compared her portion to the brother's portion and was upset that the brother's cost share was only about 8 percent (rounded up from 7.5 percent) of his monthly income. With this realization, Roberto—and many of the other students he started to convince—began to work with the professor sister's proposal. Roberto had provided the evidence that many students needed to shift their position, and there was a buzz of excitement as students began to rework their mathematical evidence. We saw many students build on Roberto's lead and apply the same strategy—comparing party cost share to monthly income—with the professor sister's proposal. For example:

**Professor Sister's Way:**

| | | |
|---|---|---|
| Professor Sister's share | 900 / 6,000 | = .15 |
| Brother | 3,150 / 20,000 | = .1575 |
| Other Sister | 450 / 3,000 | = .15 |

Students recognized that in the professor sister's proposal these ratios were practically equivalent. To them, this was a new indicator of fairness since everyone was paying an *equivalent proportion* of their salary for the party. This reasoning put the professor sister's proposal in a new light. Paying *proportionally* would be more fair than everyone paying the same "absolute" dollar amount. This activity ended with student groups presenting their positions to the class, all of which agreed with the professor sister's proposal. However, there were missed opportunities to fully engage the students in reflecting upon how these experiences changed their conceptions of fairness.

## Making Sense of Fairness

From a mathematical standpoint there are various ways students engaged proportional reasoning in this problem. We provide these additional examples in Figure 2 to show the range of mathematics accessible to middle school and high school students.

However, we did not construct the task with solely mathematical goals in mind. The central social justice issue of this task is what is a "fair share." The task is designed to help students distinguish between equal and fair, and between equal and equivalent. The success of this lesson can be measured by how students' mathematical reasoning *and* sense of fairness came together to analyze this situation and promote productive discussions. As educators who challenge ourselves to critically reflect on how we are being culturally responsive—which to us means instructional practice that connects

mathematics, children's mathematical thinking, culture, language, and issues of power/social justice—we felt good about designing a mathematically rich task stemming from a real and relevant experience that appealed to students' sense of justice.

But this activity, mathematically rich as it is, goes only so far. For example, though income disparities among family members are explicit in the task, issues of power are not. Our social justice goals for the task were only partially met. On reflection, we should have done more to explicitly engage students to develop their critical knowledge. This would include asking them to reflect on their ideas of fairness, equal, and equivalent before, during, and after doing this activity. For example, we could ask students to respond to the following prompts: *What does it mean to have a fair share? Do we think of fair share in absolute terms or in relative terms? Why? Is equal always fair? Do we consider fairness from a mathematical standpoint, social standpoint, or both?*

In terms of making an explicit stance to students about issues of fairness, our decision to gently prompt students to consider other options steered students to rethink their positions. It started with Roberto's exchange with Ms. Julia. But getting students to justify their claims was key to helping them rethink their positions about fairness from an absolute to a relative perspective. By structuring two sides in the task we invited a debate, but we could improve how we facilitated the debate to attend to the social justice issues more explicitly.

Tasks such as Papi's birthday provoke proportional reasoning and use of ratios while analyzing issues of fairness. By itself it is a beginning point, and we would argue that students need more opportunities to engage with such problems to deepen their mathematical understanding and strengthen their critical/social justice knowledge. This would include other ratio/proportional situations, such as examining racial profiling claims in traffic stops or school suspensions, fair housing prices and gentrification, or

**FIGURE 2:** OTHER SOLUTIONS INVOLVING MULTIPLICATIVE REASONING IN THE PAPI'S BIRTHDAY TASK

---

## PAPI'S BIRTHDAY TASK

Ratio of salary to total party cost: Arguing that the brother should pay it all

| **Professor Sister:** | **Brother** | **Other Sister** |
|---|---|---|
| 6,000 / 4,500 = 1.333 | 20,000 / 4,500 = 4.444 | 3,000 / 4,500 = .666 |

The brother's salary is more than four times what the party will cost, so he can afford to pay for all of it.

---

Ratio of brother's salary to each sister's salary: Arguing that the brother should pay it all

Brother to Professor Sister: 20,000 / 6,000 = 3.33

Brother to Other Sister: 20,000 / 3,000 = 6.66

Since the brother makes more than three times what the professor sister makes, and more than six times what his other sister makes, then he should pay for the whole party.

---

Finding the scale factor that determines what portion of their salary each sibling should pay: Arguing everyone should pay for the party with consideration of their income level.*

*Let X be the "operator" or multiplier to calculate the proportion each sibling will contribute from their salary. Then:*

$$20,000X + 6,000X + 3,000X = 4,500$$

$$29,000X = 4,500$$

$$X = 4,500 / 29,000$$

$$X = .1552$$

Brother pays: .1552 (20,000) = 3,104

Professor sister pays: .1552 (6,000) = 931

Other sister pays: .1552 (3,000) = 465

$$3,104 + 931 + 465 = 4,500$$

*Seeing ratio as an "operator" or multiplier that transforms (e.g., stretches or shrinks another value) is an important mathematical concept that can help to cement the multiplicative thinking we want students to utilize. Although no students in our math classes came up with this solution, we want to offer it, particularly for high school lessons. This mathematical linkage is a more advanced understanding of ratio that takes time and many experiences to develop.

---

the ratio of liquor stores (vs. social services) to residents in a neighborhood (Gutstein, 2006; Brantlinger, 2013).

We want students to discuss the merits of solutions, using mathematical evidence and providing justifications (mathematical and social) for their claims. Papi's birthday is a task that moves students to make mathematical progress that conventional proportional reasoning tasks failed to do. Future iterations of this task could include more explicit discussions about fairness, equality, and fair share, providing important experiences to deepen their mathematical knowledge and prepare them to engage as adults in current debates regarding tax structures, funding structures, and other socioeconomic political debates. This is what we strive for as social justice educators: for students to utilize mathematics as an analytical tool to understand and take action to transform their world. ☐

## REFERENCES

Brantlinger, A. (2013). Geometry of inequality. In E. Gutstein and B. Peterson (Eds.) *Rethinking mathematics: teaching social justice by the numbers, 2nd edition*. Milwaukee: Rethinking Schools. pp. 169-173.

Gutstein, E. (2006). *Reading and writing the world with mathematics: Toward a pedagogy for social justice*. New York: Routledge.

Lamon, S. J. (1999). *Teaching fractions and ratios for understanding: Essential content knowledge and instructional strategies for teachers*. Mahwah, NJ: Lawrence Erlbaum Associates.

# CHAPTER FIFTEEN

# Racism and Stop and Frisk

BY KATHRYN HIMMELSTEIN

In August 2006, Nicholas K. Peart was minding his own business, sitting with a cousin and a friend on a bench on Broadway in New York City. Suddenly, he was surrounded by NYPD squad cars. In moments, Peart found himself on the ground with a gun pointed at him. It was his 18th birthday.

Students and I read the gripping *New York Times* essay about this and several other encounters that Peart, who is African American, had with New York police. The essay became a prompt for students' own writing about a time when they, or someone they knew, had an interaction with police. I asked my students to describe: "What happened? How did you (and/or the other person) feel about it? What would you want other people to know about what happened?" I did not require students to share their narratives aloud, but I asked them to hold on to their writing throughout the unit to use as a resource. Most students were eager to have me read their narratives in class and many retold their stories to me and to other students throughout subsequent class periods.

This activity set the stage for seeing our mathematical research as intimately connected to students' lives, while also providing a valuable source of "qualitative data" for use in later activities. It was part of a project in which students used various statistical techniques (drawn from the New York State Learning Standards) to analyze publicly available data on the New York Police Department's controversial "stop, question, and frisk" policy. After students shared stories about their interactions with police, they generated a list of questions about stop and frisk, some of which are shown on the poster below.

Over the course of several weeks, we then answered these questions. We used measures of central tendency to compare stops in different New York City boroughs, created frequency histograms to examine which ages were stopped the most, and produced cumulative frequency histograms to analyze seasonal variations in stops. We completed the activity detailed below near the end of the unit, after students practiced calculating five-number summaries and constructing box-and-whisker plots using smaller data sets involving lower numbers. In this activity, I asked students to combine both qualitative data, including their previously created personal narratives, with quantitative data, specifically box-and-whisker plots. This highlights the value of using both qualitative and quantitative data in constructing a persuasive argument.

Following this lesson, we returned to the "Stop, Question, and Frisk in New York Neighborhoods" graphic and the "Brooklyn Residents" table included in students' handouts (shown in the following pages). Our discussion revealed a major flaw in the arguments students had crafted during the lesson below: a valid analysis of racial disparities in stop and frisk would need to consider not the *number* of stops of people of different races, but rather the *rate* of stops for different racial groups relative to their representation in the general population. To address this problem, students calculated the "representation ratios" of the four racial groups for each borough by dividing the group's share of stops in the borough (calculated as a percentage by summing the numbers from each precinct) by their share of the general population (shown in the "Brooklyn Residents" table). Students then summarized the implications of these representation ratios and explained how the information they communicate is different from the information contained in absolute numbers of stops by race.

The reason for including this exploration in a separate lesson was that the difference between relative and absolute comparisons, and how they can (or cannot) be used to demonstrate discrimination, is a subtle and challenging concept for many students. The primary aim of the lesson below was to begin to understand the role of quantitative and qualitative data in argumentation. The discussion of how an inappropriate use of absolute comparisons may be inaccurate or misleading—even when the data on which it is based is "correct" (i.e., not falsified)—was left for a future lesson. One problem that arose from separating these lessons was that some students, having invested time in constructing an argument based on the absolute numbers of stops, seemed convinced that the absolute comparison was sufficient evidence of racial discrimination and failed to understand the need for a relative comparison. To avoid this, it might be preferable to have students immediately calculate representation ratios and use these as their quantitative data in the lesson, rather than constructing box-and-whisker plots. In either configuration, the purpose of this lesson is to emphasize the important roles of both qualitative and quantitative data, affirming students' experiences and mathematical analyses as two powerful tools to document racial discrimination in stop and frisks. □

## ACTIVITY BOX

**Aim**: Students will use qualitative and quantitative data to construct arguments about racism and stop and frisk.

**Opening Question** *(posted on the board; students are given 7-10 minutes to write individual responses)*: Based on your personal experience, do you think all racial groups are targeted equally for stop and frisks? What evidence do you have to support your answer? (You may wish to refer to your personal narrative from earlier in the unit and describe a particular incident you experienced or witnessed.)

**Mini Lesson**:

1. Students share out their answers to the opening question, while the teacher records a brief outline of each story or answer on the board. Note that students are providing *qualitative* data to support the claim that people may be targeted for stop and frisks based on their race.

2. The teacher writes a definition of qualitative data on the board noting that it involves description and is observed, not measured. Students generate examples of qualitative data about themselves, and record these (examples include: friendly, motivated, positive). Then contrast qualitative data with quantitative data, recording a definition on the board and noting that quantitative data involves measurement or amount. It may be useful to have students make a connection to online shopping or grocery store receipts that refer to quantity. Have students generate examples of quantitative data about themselves, and record these (examples include: height, weight, age).

3. Note that if people claim that stop and frisks are—or are not—conducted in a racially discriminatory manner, then they use both qualitative and quantitative data to support their argument. Discuss why this is: What are the advantages and disadvantages of the qualitative data we have already presented on the board? How could using quantitative data contribute to our argument?

## ACTIVITY BOX (CONTINUED)

4. Have students read aloud their question for the day (projected or written on the board and printed at the top of their handouts): "Is there racial discrimination in who is stopped and frisked?" Student answers should be backed up with the following evidence:

- Box-and-whisker plots comparing the number of black, Asian, Latina/o, and white people stopped and frisked in Brooklyn precincts in the last quarter of 2011.

- At least one more piece of quantitative data drawn from the "Six N.Y. Lawmakers Wear Hoodies in Honor of Fla. Teen" article; the "Stop, Question, and Frisk in New York Neighborhoods" graphic; or the "Brooklyn Residents" table.

- At least one piece of qualitative data drawn from the articles listed above or from your own narratives.

**Activity**: *See student handout (page 126).*

**Summary**:

1. Post students' posters at the front of the room.

2. Discuss: Is there racial discrimination in who is stopped and frisked? What qualitative data do we have that there is or is not discrimination? What quantitative data do we have that there is or is not discrimination? Make sure to discuss how stacked box-and-whisker plots (as opposed to unstacked) make it easier to compare the number of stops by race.

3. Discuss: What are the advantages and disadvantages of using qualitative data to support our argument? What are the advantages and disadvantages of using quantitative data to support our argument? Who might be convinced by each type of data? Who might be skeptical of each type of data? What are the advantages of using both types of data, rather than just one type?

**Final Questions** *(posted on the board)*: What is the difference between qualitative and quantitative data? What is the purpose of using both types of data to support an argument?

# STUDENT HANDOUT

## RACE AND STOP-AND-FRISKS POSTER

### Is there racial discrimination in who is stopped and frisked? How do you know? What is the evidence?

Required Element #1: A box-and-whisker plot comparing the number of black, Asian, Latina/o, and white people stopped in different Brooklyn precincts.

On a piece of graph paper, construct four stacked box-and-whisker plots to compare the stop-and-frisks of black, Asian, Latina/o, and white people by precincts in Brooklyn.

*[The four tables are separate here because students would otherwise need to put the number of stops in ascending order to create the five-number summaries and the box-and-whisker plots. The tradeoff is that this representation makes it harder to compare the number of stops by race within precincts.]*

| Precinct | Number of Latina/o people stopped and frisked | Precinct | Number of black people stopped and frisked | Precinct | Number of white people stopped and frisked | Precinct | Number of Asian people stopped and frisked |
|---|---|---|---|---|---|---|---|
| 71 | 86 | 68 | 47 | 67 | 18 | 88 | 8 |
| 67 | 100 | 62 | 48 | 71 | 20 | 71 | 9 |
| 69 | 121 | 72 | 90 | 81 | 23 | 69 | 10 |
| 63 | 130 | 66 | 137 | 73 | 33 | 67 | 11 |
| 77 | 131 | 94 | 139 | 69 | 37 | 94 | 12 |
| 94 | 143 | 61 | 332 | 77 | 45 | 73 | 14 |
| 79 | 161 | 76 | 342 | 75 | 48 | 77 | 18 |
| 73 | 167 | 78 | 350 | 79 | 68 | 81 | 18 |
| 81 | 185 | 60 | 691 | 88 | 76 | 63 | 20 |
| 84 | 206 | 84 | 715 | 72 | 91 | 76 | 21 |
| 88 | 211 | 70 | 858 | 83 | 103 | 79 | 21 |
| 70 | 232 | 83 | 864 | 94 | 121 | 78 | 22 |
| 62 | 235 | 90 | 874 | 70 | 134 | 68 | 25 |
| 68 | 247 | 71 | 898 | 84 | 144 | 83 | 30 |
| 76 | 275 | 63 | 998 | 63 | 160 | 90 | 45 |
| 61 | 292 | 79 | 1,006 | 76 | 161 | 75 | 48 |
| 78 | 329 | 73 | 1,253 | 78 | 178 | 62 | 52 |
| 60 | 456 | 69 | 1,286 | 66 | 220 | 61 | 60 |
| 66 | 741 | 88 | 1,317 | 90 | 325 | 72 | 67 |
| 75 | 746 | 77 | 1,351 | 62 | 370 | 60 | 70 |
| 83 | 1,065 | 81 | 1,380 | 68 | 418 | 84 | 70 |
| 72 | 1,159 | 67 | 1,878 | 60 | 509 | 70 | 76 |
| 90 | 1,586 | 75 | 2,113 | 61 | 694 | 66 | 192 |

*Source: http://www.nyclu.org/files/releases/2011_4th_Qrtr_report.pdf*

126

**RACE AND STOP-AND-FRISKS POSTER** (continued)

**REQUIRED ELEMENT #2: <u>AT LEAST</u> ONE ADDITIONAL PIECE OF QUANTITATIVE DATA**

This may be taken from the tables above, or from any of the sources on the back of this sheet. Make sure to select a piece of data that illustrates or supports your answer to the question.

**REQUIRED ELEMENT #3: <u>AT LEAST</u> ONE PIECE OF QUALITATIVE DATA**

This may be taken from any of the sources on the back of this sheet, or from your group members' narratives describing their experiences. Make sure to select a piece of data that illustrates or supports your answer to the question.

---

### 6 NY LAWMAKERS WEAR HOODIES IN HONOR OF FLA. TEEN

By Michael Gormley
March 26, 2012

ALBANY, N.Y. (AP)—Six New York state senators wore "hoodies" in the staid, 235-year-old Capitol chamber to express their outrage at the deadly shooting of a black youth in Florida that they blame on public attitudes born in New York City.

The Democrats from the city, four of them black and two white, wore their gray and blue suit jackets over their hoodies, remaining in conformance with the Senate rules in the 62-seat chamber. There was no comment from the Republican majority.

"It was born here in New York City and now it has cascaded all the way down to the southern coast of Florida," said Sen. Eric Adams, a Democrat and former sergeant with the New York City Police Department. "The stop-and-frisk policy gave birth not only to police officers believing that a person of color is automatically a criminal, now it has grown into the civilian patrol units."

Trayvon Martin, 17, died in Florida one month ago on Monday, shot by a neighborhood watch volunteer who has not been charged. Martin wore a hoodie as he walked home on a rainy night in a gated community, carrying only his cell phone, an iced tea container, and Skittles candy.

On Monday, the senators recited a list of black men from New York recently killed in police confrontations and blamed a crackdown on crime dating to the mayoral administration of Rudy Guiliani in the 1990s. That includes community policing that involves questioning residents.

Since then, crime in the city has dropped to historic lows, including a drop in murders by half in a decade, and Mayor Michael Bloomberg refers to New York as the safest big city in the world.

New York City police made a record 684,330 street stops last year, and 87 percent of those targeted were either black or Hispanic under a policy that has been vilified by civil rights groups, but lauded as an essential crime-fighting and life-saving tool by department officials.

---

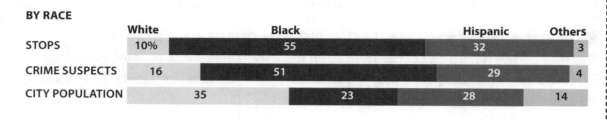

## STOP, QUESTION, AND FRISK IN NEW YORK NEIGHBORHOODS
By Matthew Bloch, Ford Fessenden, and Janet Roberts

*The New York Times*, July 10, 2010

**BY RACE**

| | White | Black | Hispanic | Others |
|---|---|---|---|---|
| STOPS | 10% | 55 | 32 | 3 |
| CRIME SUSPECTS | 16 | 51 | 29 | 4 |
| CITY POPULATION | 35 | 23 | 28 | 14 |

*Source: http://www.nytimes.com/interactive/2010/07/11/nyregion/20100711-stop-and-frisk.html?ref=nyregion*

## BROOKLYN RESIDENTS
### U.S. CENSUS BUREAU

| Asian | 10 percent |
|---|---|
| Black | 34 percent |
| White | 36 percent |
| Latina/o | 20 percent |

*Source: http://quickfacts.census.gov/qfd/states/36/36047.html*

## EXAMPLES OF STUDENT WORK

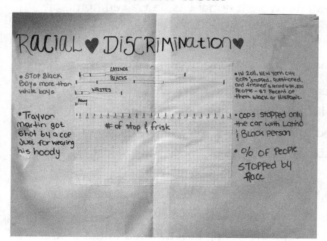

# CHAPTER SIXTEEN

# "With Math, It's Like You Have More Defense"

## Students Investigate Overcrowding at Their School

BY ERIN E. TURNER AND
BEATRIZ T. FONT STRAWHUN

In an overcrowded New York middle school, students discovered that math was a path to investigating and working to change conditions at their school.

After planning a unit with Erin, a university professor, Beatriz taught a six-week unit where students used mathematics to investigate issues of overcrowding at Francis Middle School, which is located in a predominantly working class African American, Dominican, and Puerto Rican community in New York City. Erin was present in the classroom for all of the class discussions.

When the school opened in 1990, following a district call for the creation of more "small schools" for middle-grade students, the founding (and current) principal faced the near impossible task of finding available space for the school within the district. At that time, the elementary school that currently shares a building with Francis had vacated the top floor because of continual problems with a leaking roof. Francis' principal claimed the space for the new school, repairs began, and the school opened shortly thereafter.

Although Francis began as a small school, districtwide changes in student enrollment caused the school's student population to grow dramatically, from approximately 145 students to 213 students, and it was projected to grow another 15 to 20 percent (an additional 30 to 40 students) the following year. So as students climbed the five flights of stairs to reach the once pigeon-infested fifth floor that now housed their school, they worried about how the narrow halls and lack of classroom space might accommodate their already cramped school community. Some students were concerned about the potential fire hazards created by the school's long, narrow hallways. A building across the street had recently been damaged in a fire, and in the aftermath of 9/11, students were all too aware of the dangers of being trapped in a burning building. Others felt their classrooms, several of which had recently been subdivided to create multiple rooms, were too small and overcrowded. Floor-to-ceiling columns and smaller poles scattered throughout classrooms made simple tasks like seeing the chalkboard difficult.

It was students' concerns about the lack of space at their school that guided the initial development of this project. One of our primary goals was to design a unit of study that drew upon students' interests and experiences and provided students opportunities to learn and use mathematics in personally and socially meaningful ways. With this in mind, Beatriz asked students to list issues about the school and local community that concerned them.

Several topics appeared repeatedly in students' lists, including violence in the neighborhood; health issues such as AIDS; racism and sexism in the media; and the "space crisis" at the school. While any one of these topics might have sparked a rich mathematical unit, we selected the issue of "Overcrowding at Our School" for several reasons, including (a) the rich mathematical content the unit would draw upon, (b) the opportunities it would provide for students to generate their own data versus analyzing data from an external source, (c) the salience this issue had for students, and (d) the potential links to issues of equity and fairness.

Initially, students claimed they were more crowded than other schools and were eager to speak out in hopes of increasing their school space. Students were particularly bothered by the disparity they observed between their own school space and that of Longmore, another small middle school that had recently moved into the fourth floor of the same school building. (Note: When the fourth floor initially became available, Francis lobbied to move into it, but their request was not granted, and the space was instead awarded to Longmore, a technology magnet school that attracted affluent, predominately white families from across the district.)

Yet they were not sure how to talk about the crowding in terms that might convince others, and it was unclear to them how mathematics could support their argument. To help students connect their concerns about overcrowding with mathematical tools that would support their investigation, Beatriz posed questions such as: How can we show them how much space we have? What kind of information would we need to collect? What kinds of measurements should we make? How can we prove that we are more crowded than Longmore? Students quickly realized that quantifying the school's space would be helpful.

One student, Jhana, raised a concern about the tight space in the hallways after second period, a time when all 213 students in the school were simultaneously released from their classrooms. In class, she repeatedly lobbied to investigate this. "What we need to know," she argued during a discussion, "is after second period, because that's when the most kids come out. How many kids get dismissed? ... And [we need] the area of all the hallways." Other students agreed that finding the area of classroom and hallway spaces would "give proof" to their claims of overcrowding. So Beatriz prepared a series of mini-lessons that addressed concepts such as linear measurement and how to find the area of spaces with mixed dimensions, such as a hallway that measured 10-1/2 meters by 1-1/4 meters.

Jhana worked with several class-mates to measure and calculate the area of the school's hallways and used ratios to compare the hallway space per student in her school, Francis, to the hallway space per student at Longmore. She argued that ratios "make it easier to see the big difference" and noted:

> [Before] I wouldn't really use math. I would just say, LOOK how much space they have [in their school] instead of what we have [in our school]. ... But I would really use math now. ... Math made my argument make more sense, and have more of an idea, and actually tell what is happening, because it gave more detail to it.

As the class continued to analyze overcrowding at their school, they discovered disparities between their own space and that of other schools, and numerous instances where their school violated district building codes.

For example, during one lesson after students had worked in teams to measure and calculate the area of different classrooms and hallways, Beatriz asked each group to share their measurements. Students were shocked as they viewed the size of their own classrooms (e.g., 474 square feet, 497 square feet, 567 square feet) compared to the classrooms at Longmore (e.g., 772 square feet, 864 square feet, 918 square feet). One student commented, "It's not fair! They have a smaller amount of students and bigger classrooms. They have to keep cutting our classrooms in half because we have so many kids."

Ultimately, students decided to share this information with the district. They wrote letters to the superintendent, prepared fact sheets with the results of their analysis for administrators, and spoke at a school governing board meeting. As Jhana considered whether the class's analysis of the space crisis at their school made a difference, she commented: "Yes [we made a difference], because first of all, we found out something for ourselves, and we actually proved a point. ... WE made the

## Students' participation in this unit helped develop their senses of themselves as people who make a difference.

difference. ... We learned what we learned, and we told people. ... Math made our argument make more sense. .... You couldn't do it without the math."

As mathematics educators, we would like all of our students to exhibit such passion about the power of mathematics. Jhana and her class-mates invented novel problem-solving strategies and used mathematics as a tool to analyze and act upon situations at their school. We believe that students' participation in this unit helped develop their senses of themselves as people who make a difference.

### Students Negotiate the Curriculum

Throughout the "Overcrowding at Our School" project, the students had opportunities to insert their interests, goals, and purposes into the curriculum. For example, after several days of measuring classrooms and calculating areas, students formed small groups to pose their own problem about a particular aspect of the school space.

Beatriz asked students to identify one issue dealing with overcrowding at the school and to discuss how they might use mathematics to find out more about the situation. As students posed problems that mattered to them, their desire to understand and affect the overcrowding increased their engagement in mathematics, and thereby enhanced the learning that occurred.

Angel, a tall and rather quiet African American student, was not a frequent participant in problem-solving discussions before this unit. But when the class began to investigate overcrowding at their school, there was a notable shift in Angel's level of engagement. Angel was extremely concerned about the school's bathrooms. She found it difficult to navigate among the other 10 or 12 people in the tight space. She noted that all females in the school, including 103 students and 15 teachers, had to share one rather small facility with only three working stalls, and a very small sink station. So when Beatriz asked Angel's group what aspect of the school space they wanted to investigate, the choice for Angel was obvious: "We want to know: Why are the girls' bathrooms so small?"

Angel's group constructed a floor plan of the restroom, measured its dimensions, calculated the area, and then analyzed the bathroom space based on the number of stalls, the estimated wait time during peak use periods, and the space available for waiting. Angel spoke about how the opportunity to investigate an issue she cared about made her feel "mad curious" and drew her into the mathematics. She commented:

> It was easier to do the math this way, instead of just learning it straight, like solving a problem, because we would actually, like, really get into it, and that made it easier. … Like the facts [about the school], they made you want to find out the answer. Like we wanted to know."

For other students, the opportunity to investigate real issues not only increased their engagement, but also pushed them to construct and apply important mathematical concepts.

Lianna, like Jhana, was concerned about the school's narrow and densely populated hallways. As she left Beatriz's classroom each day, she faced the continual challenge of navigating through one of the school's narrowest and most densely populated hallways. Not a student who was comfortable pushing her way through oncoming crowds of up to 80 children at a time, Lianna was often left standing just outside the door for four or five minutes while other students passed in and out of adjacent classrooms. Her group decided to compare the total hallway area at Francis with the area of the hallways of Longmore. The mathematics that Lianna's group engaged in would qualify as rigorous in any 6th-grade classroom. They developed their own strategies for multiplying mixed numbers to find the area of hallways with dimensions like 18-3/4 by 1-1/4 meters, and subdivided irregular spaces into rectangular and triangular areas.

But Lianna was not content with simply stating the total hallway area of each school; she wanted to make her argument stronger or, to use her words, to "use more specifics so people will listen." When she overheard that a classmate, Thomas, had calculated the ratio of hallway space per student, she was intrigued.

"How did you do that?" she asked. "We already found out the [hallway] area of Longmore, and I want to see how much [space] they will each get. You found out how much each person will get in Francis, and I want to do the same thing in Longmore. But I don't know how to do it."

"You've got to know how many students there are," said Thomas.

"Sixty," said Lianna.

"Sixty students. And how much is the area?" asked Thomas.

"Two hundred forty-six and 3/4 meters squared," she answered.

"So I am going to divide 60 into 246," said Thomas. "Because that way I can find out how much each person gets, cause it kind of divides it [the space] up.

Several days before this conversation, Beatriz had presented a mini-lesson designed to help students think about overcrowding in terms of "space-to-people" ratios. Many of the students found that comparing ratios helped support their claims of overcrowding, and some of them, like Thomas, had become adept at using this mathematical tool. It was not uncommon for students in the classroom to ask each other for help or to question each other's strategies in order to understand them better, as Lianna did. With Thomas's help, she figured out that if all students at Longmore entered the hallways at the same time, each student would have 4.1 square meters to her/himself. She was shocked when she compared that figure to the less than one square meter of hallway space per student at her school.

## Taking Action

We wanted to support students in sharing what they learned through their investigations with the school and neighborhood community. Beatriz helped students to brainstorm ways that they might educate others about overcrowding at the school. Students generated lots of ideas, including distributing flyers, visiting the school board, going "on strike," making a large floor plan of the school to display, and compiling all their data to share with the district. Except for going on strike, the students implemented all of these ideas.

At the end of the unit, Naisha, a spirited and opinionated African American student, spoke at a school advisory council meeting. This council represented the school at the district level and helped make decisions on matters of spending, curriculum and assessment, staffing, and enrollment. Naisha embraced the opportunity

and volunteered to prepare a speech. What follows is the text of the speech she presented to the board:

> Good evening, my name is Naisha Watson. I am a 6th grader at Francis Middle School, and I am going to talk about overcrowding at our school. Our math class has been comparing our school

"The 3,000 participants at the [2002] World Economic Forum, which drifted through the hallways of the Waldorf Astoria, dropped $100 million on New York hotels, ballrooms, and restaurants, according to the New York City tourism board. That comes out to $33,333.33 per person. In five days in New York, each participant of the World Economic Forum spent on average what the average American makes in a year, four times what the average Mexican makes in a year, 14 times what the average person in India makes in a year, 22 times what the average person makes in Bangladesh, and 74 times what the average person makes in a year in Sierra Leone, according to United Nations figures."

**— Derrick Z. Jackson, "The Elite's Pure Greed,"**
*The Boston Globe*, Feb. 8, 2002

> to Longmore. We have noticed, as a class, that we have no space for kids to sit. … The board of education has a building code that the classrooms have to be at least 750 square feet for 30 children. As you can see on the graph, only 3 classrooms are big enough, the rest of the classrooms that are orange on the graph are smaller than 750 square feet. [Refers to a large diagram of the school created by several students.] … The board of education has another building code that says the hallways must be 5 feet and 8 inches wide. … There is only one hallway that is 5 feet and 8 inches. All the other hallways that are red on the graph do not meet the board of education building code. … In our school we have 213

kids. If there was a fire in our school it would be a hazard to get through our narrow halls. So, as a school, we think we should have less students or more space.

Naisha felt that speaking out as a way of resisting the inequities her class discovered was not only necessary, but also potentially effective. "I think it's good [that we talked to the district], because if you keep talking to them then they will probably listen," she explained. "And you will get on their nerves and maybe then they will want to give us more space, or let us be in a different building with more space, that is lawful."

Even though the students knew that the district lacked funds to build a new school or add a floor to the building, they felt good about contributing to the public discussion of overcrowding at the school. As Lianna argued, "We have to say something because we are the students and we are the ones that have to live in the school every day."

### Challenging Students' Ideas

Not only did opportunities to engage in responsive action support students' sense of themselves as people who can and do make a difference, but using mathematics as a tool to support their actions challenged students' view of the discipline. For instance, when we initially asked Naisha what she thought of mathematics, she responded, "What do I think of math—you

## The opportunity to investigate real issues pushed students to construct and apply important mathematical concepts.

mean, numbers?" She described math as something she felt good about only when she got the answers right.

In contrast, as Naisha reflected on Beatriz's class at the end of the semester, she explained that unlike previous classes, where she studied material but never had the opportunity to "do anything" with what she learned, in this class, "we did something with it. ... Without the math, then, we wouldn't have the area of the school, and we wouldn't really know. And the [district] meeting wouldn't have been as powerful as it was."

Naisha was not the only student who began to recognize that math "made [her] arguments make more sense." Other students said math helped them to "prove how most stuff is not shared evenly," and "to prove to the district that our school was smaller;" math "gave more details" and "specifics" to their arguments and afforded them "more defense" in the problems they were fighting against. Students also spoke eloquently about how they drew upon mathematics to address "things in [the] community and school," and referred to this way of engaging with the discipline as "a lifelong thing" that is not only about mathematics but also about "things that you be in everyday, and it's a part of your life." Given that students often struggle to identify personal reasons why they should learn mathematics, these shifts in their understanding of the discipline are significant.

### Further Reflections

As we reflected on the project, we found that creating space for students to pose their own problems and to inject their interests and concerns into the curriculum was a powerful way of supporting student activism. Occasionally, students posed problems about the school that did not lend themselves to rich mathematical investigations. We recognize that teachers have a responsibility to ensure that students learn certain content, and students' interests may not always lead them to a given mathematical idea. Beatriz had clear mathematical goals in mind for this unit (linear and area measurement, ratio, operations with fractions, and mixed numbers). But she thought it was equally important that students participated in mathematics projects that were personally and socially meaningful. We acknowledge that mathematics may not always be the best discipline to address the questions that students pose. Beatriz's challenge was to work with students to negotiate an intersection between their interests and the mathematical content they needed to study.

We also found that creating a classroom culture where critique was welcomed, and even expected, was essential. It was important for students to feel safe posing difficult questions, such as those that alluded to the connection between their particular situation at Francis (a poor school with inadequate space and resources), and broader educational inequalities that exist along the lines of race and class. While classroom discussions began with matter-of-fact statements about the data students were collecting and what they disliked about their school space, through questions such as, "Why do you think it is like that?" the talk gradually shifted to an exploration of why there were particular discrepancies, namely those between their own neighborhood school and Longmore. Students argued that the superior conditions at Longmore (a more affluent school) were not random but were directly related to the race and socioeconomic status of the students. "There are more white people than anyone else at that school," one student noted, and "the white people always get the good education, it's like an upper-class thing, for the white kids. ... That's just how it is!" In further discussion, students suggested that the demographics (in particular the socioeconomic status) of Longmore students were unlike those of Francis, and that these differences in demographics might be linked to the discrepancies in both the condition and size of school facilities, and to the "protection" of the school by those in power.

Beatriz opened a space in her classroom for students to approach their situation with a critical mindset, and in doing so, she supported their sense that they can act and make a difference.

At the end of the study, it was still unclear whether the district would increase the school's allocated space or make any adjustment to the number of incoming students that Francis had to accept. So unfortunately, the students ended 6th grade not knowing whether their efforts had any direct impact. But over the summer, the district decided to reduce Francis's incoming class by approximately 30 students, which allowed the school to retain its current size instead of increasing from 213 to 240 students, as initially planned. The district's action prevented an already overcrowded school from growing, which students welcomed as one small success.

Yet, what seems most important is not whether this particular "battle" was won or lost, but the shifts in understanding and increased critical awareness that students took from the experience. As Jhana put it: "We found out something for ourselves, and we actually proved a point. We made a difference. Math made our argument make more sense. You couldn't do it without the math." □

All students' names have been changed.

# DESIGNING A WHEELCHAIR RAMP

## PUTTING THE PYTHAGOREAN THEOREM TO USE

BY ADAM RENNER

As a geometry teacher and community-service-club moderator, I constantly sought more critical and authentic ways to teach secondary math. Taking a project-based focus by my third year, I also wanted to connect math to other disciplines and to issues of social justice.

One project involved the design of a community garden: Students designed scale drawings, estimated costs, and investigated issues of hunger and poverty in the local and global communities.

Another example involved designing a wheelchair ramp on campus. Since ours was an old, privately run school, we had somehow avoided mandates to make the building handicapped-accessible. While the building did have an elevator, the only ground-level entrance was a door inconveniently located in the basement, at the rear of the school. When it came time to teach the Pythagorean theorem during my second year, I decided to have the students design a wheelchair ramp. Students were required to figure out the best place to put a wheelchair ramp, develop a scale drawing, and contact a local contractor to find out the building cost based on their measurements and calculations. Measuring the height of the steps and knowing that for every one foot of rise, 12 feet of run were required, students would have to use the Pythagorean theorem to calculate the length of the ramp: the hypotenuse.

### THE FOLLOW-UP

When students presented their designs and the contractors' bids for the wheelchair ramp to the school's administration, the students were thanked for their hard work, but their suggestions for building the ramp ultimately were ignored. Our follow-up discussion in class fruitfully centered on issues of public versus private facilities, the federal Americans with Disabilities Act, the fact that no students in this school were, at that time, disabled, and the apparent budgetary priorities of the school.

During my third year of teaching, when it came time for the Pythagorean theorem, students knew the administration previously had dismissed suggestions that a contractor build a wheelchair ramp. This time, the students offered to build the ramp themselves as a community-service-club project. Once again, the students were thanked for their work, but the plan was rejected. We discussed the same issues the students had covered during the previous year.

Sadly, later that year, a student was involved in a car accident which left her in a wheelchair. Watching her enter the school, day after day, via the rear of the school was a sad reminder of the denial of the students' work.

### FIVE YEARS LATER

Five years later, after I had left the school, there was still no wheelchair ramp. As a more experienced teacher, I now wonder how I might have handled the situation differently. Should we have attempted to garner more support from others at the school? Should we have been more visible in our opposition to the administration's decision? Should we have sought legal help? I will always wonder if I missed an opportunity. ∎

# CHAPTER SEVENTEEN

# 'That Smell Is from the Sugar Cane Fire!'

BY RAQUEL MILANI AND
MICHELA TUCHAPESK DA SILVA

Leafing through a school mathematics book in Brazil, where we live and work, we found a picture of the Amazon Rainforest's trees being burned. This national problem reminded us of a regional problem: the sugar cane fires in the interior of the state of São Paulo. This state is Brazil's largest producer of sugar cane, and fires are commonplace in the harvest season. Its serious consequences impact the everyday lives of the people who live in that region.

In homes, on streets, and in classrooms, one can smell the strong odor from the sugar cane fires. Teachers can transform this feeling into an activity for mathematics classes. Students could start by reading news reports and address questions such as: Why do fires happen in sugar cane fields? What are alternative strategies to harvest the sugar? What damage do the fires cause to soil, fauna, flora, and people? What are the existing laws on fires? Teachers and students can then use mathematics to problematize these questions. In this short piece, we give some examples of what teachers could do in Brazil and through these, hope to spark others to think about how they might do similar work in their own locale.

> Sugar cane harvesting begins during the state of São Paulo's dry season. Fire is the harvest method used in 45 percent of the 4.7 million hectares cultivated in the state. Machines harvest the remaining area.

In order to help students get a sense of space about the burned region, a teacher could ask students to geometrically represent the region (for example, show 45 percent of a rectangle), calculate the burned area in hectares, convert hectares to square meters, and compare the area with the area of the neighborhood where the school is located, or even the city where students live. When students visualize a familiar region and compare it with the burned area, they may have a clearer idea about the situation. We believe that the image is stronger than a number, and the information may cause students to be concerned.

A possible alternative to reduce fire damage to the environment is to mechanize the harvest. However, even with technological advances, many companies choose manual labor in areas where fires are still allowed. Why do companies do this? Is it cheaper than mechanical harvesting? These questions are common to residents and arise every fire season. There are laws that regulate how long a fire can burn. The machines can harvest sugar cane planted in relatively flat lands (an incline of no more than 12 percent), but even on steeper land, the law still specifies how long the fires can burn. Until the fires burn out, should residents have to breathe polluted air? Teacher and students can investigate the costs to the companies of mechanization as compared to manual harvesting to help understand decisions that may save companies money at the cost of the people's health.

Students can then find out which municipality charges the highest fine and question why there is a difference between cities. Once the city collects the fine, where does it go? Does it benefit the environment? The laws were created because the fires damage the environment, cause air pollution, reduce productivity of various crops, and also harm human health, such as causing respiratory diseases. Research reveals how dangerous the fires are to residents' and workers' health.

> In Araraquara, a city in the state of São Paulo, the fine is 346 reais (about $168) for every 500 m² of burned area. If the fire exceeds this limit, an additional fee of 173 reais (about $84) is added to the value for each 250 m² burned. The city of São Carlos charges 0.96 reais (about $.47) per square meter of burned area. The city reported 213 owners of land for fire violations.

The sugar harvesters' lives are precarious. They are called *boias-frias* (boia is an informal word for food; fria means cold), which comes from the fact that workers bring their own food, cooked early in the morning, and eat it cold in the plantation. With the fires, they are exposed to temperatures of about 45° C (about 113° F). They also run the risk of cutting themselves with sharp work instruments and being attacked by poisonous animals that inhabit the plantations. Sugar cane soot penetrates the body through the skin and lungs and circulates in workers' bloodstreams, causing diseases.

A recent study from the University of São Carlos points out that in the state of São Paulo, each worker cuts an average of 12 [Brazilian] tons (12,000 kilograms) per day in exchange for 24 reais (about $11.66) and walks about 9 km (about 5.6 miles), strikes 72,000 machete blows, and carries 800 bundles of 15 kg of sugar cane. According to the study, five of every 100 accidents involving workers with a formal contract occur in the sugar-alcohol industry. From 2002 to 2005, there were 83,000 accidents with 312 deaths in the production chain of sugar and alcohol.

With these data, students can estimate how much money those workers earn per hour (if the workday is eight or 10 hours, for example) and per month. A comparison between the monthly cost of food and the salary of sugar cane harvesters can spark conversations about those workers' quality of life, and the concept of proportion can help students predict the number of future accidents. Teachers can pose questions about the relationship between the distance workers walk in a day and the physical effort expended. These calculations and discussions can help teachers and students reflect on the working conditions of sugar cane workers, and teachers could also have students investigate the relationship between the payroll of the sugar-alcohol companies and their profits. Also, in cities close to where fires are concentrated, soot covers the yards, and residents must wash this area frequently. People complain that their water bill increases due to the cleaning. Students can compare the bills before and during the fires and reflect on the cost differences.

Behind the strong odor caused by the sugar cane fires, dirt, soot, and respiratory diseases, there are mathematical concepts that help students and teachers more deeply understand their lives and reflect on the future of where they live and their quality of life. The context of sugar cane fires is a way for teachers and students to engage in discussions of social responsibility and justice—in math class. What is the responsibility of everyone involved (companies, government, residents)? What right do companies have to affect the very air that people breathe?

We are aware that classroom activities, such as the sugar cane fire investigation, do not necessarily cause great movements of protest against social injustices. The transformations that we seek are related to students' ways of thinking. What did students know and think about sugar cane fires before the activity? Do they think differently afterward? What will they feel when they breathe air with the odor of sugar cane fires? How will they respond to media reports about the subject? What will they tell their friends and family? In reflecting on what affects our lives as teachers and students, we proposed activities in which numbers, pictures, and graphs have something "to say." The teachers' task is to create opportunities for students to "hear" the numbers and, through this, for them to develop competencies in reading, interpretation, comparison, and reflection as necessary steps in transforming their consciousness and the world. □

# CHAPTER EIGHTEEN

# What's a Good School?

## Using Data to Define Measures of a School's Success

BY MICHELLE ALLMAN AND ALMANZIA OPEYO

I asked my math students at Champion High School, an alternative school in Brockton, Massachusetts, to pair up and identify five ways they would measure the "success" of a school. These new 9th graders began with responses that included good teachers, better lunches, fair principals, fun classrooms, extracurricular activities, and no "drama." This initial brainstorm began a three-week journey designed not only to help these students who possessed 6th-grade-level math skills develop the use of variables to create complex meaningful expressions and apply proportional reasoning to analyze real-world situations, but also—and even more importantly—to use these fundamental math skills to explore equity in our education system.

The students' preliminary responses about how to define a successful school were somewhat surprising given that Champion students seem ripe to challenge the status quo. Students at this school consciously chose to attend this alternative high school because Brockton's 4,000-student comprehensive high school did not offer what they felt they needed to interrupt their earlier patterns of disengagement from a traditional education system that so often fails to provide an environment of positive attention and support. Yet for these linguistically, racially, ethnically, and socioeconomically diverse 9th graders who are often on the wrong side of society's dividing lines, their lists showed they either did not recognize issues of equity or did not feel comfortable raising such concerns in a classroom setting. They also struggled with how to describe *measurable* attributes of a successful school. Both these matters needed to be addressed through experiences that would improve their ability to use math to interpret the world and help them to recognize and discuss issues of injustice that are undoubtedly keenly felt, yet rarely acknowledged in a classroom.

We—Michelle, their math teacher, and Almanzia, a data analyst based in Atlanta who has expertise in researching equity in schools—guided the students in this three-week project. Because of our previous connection as colleagues in an organization committed to educational reform and rooted in social justice, we had built a foundation that would allow us to guide students in difficult conversations about identity and privilege. These types of "courageous" conversations are necessary to plan and create a learning environment where students can safely examine and discuss school and societal data that could be painful, inflammatory, embarrassing, or accusatory if not well framed. The following were the key tenets underlying the three-week unit: (1) students *had* to learn the math skills and concepts that they would need for their future algebra course; (2) the math itself is virtually meaningless if students cannot apply it to real-world situations; (3) one measures a school's success not only by its aggregate data, but also by how well it serves the range of populations within the school; and (4) having students learn from and present their work to an audience outside the classroom has significant academic and personal benefits.

Students began by learning to use variables and expressions to represent the world. Starting with variables representing things they were familiar with (like the number of adults, children, books, televisions, etc., in a home) I asked students to work in groups to estimate average values for these variables and then use those values to create numerical and variable expressions that represented something meaningful. My use of this context was an unfortunate one, where I made the incorrect assumption that in a school comprised of predominantly low-income students, they would feel somehow collectively comfortable with their economic situation. So when some students took this exercise as an opportunity to boast to others about the number of shoes or televisions in their homes, I used this to discuss choices and circumstances that affect what we own. Yet it would have been wiser either to be upfront

**A student found that (M+W) x S + C x P would give the total number of shoes on average in a home where:**

**M**  The average number of Men per home (males 18 and older)

**W**  The average number of Women per home (females 18 and older)

**C**  The average number of Children per home (males and females under 18)

**S**  The average number of pairs of Shoes owned by an adult

**P**  The average number of Pairs of shoes owned by a child

about how these conversations can trigger issues around class and wealth or to have used less problematic variables such as those related to a food court at a mall or movie theater.

It was at this point that the class had their first Skype session with Almanzia, who worked with students to apply their new skills with variables and expressions to their original lists of how to measure school success and created more measurable attributes. During this session we pushed students who listed things like "good teachers," "fair principals," and "technology" to concretely define these variables and explain what they would *count* to capture these constructs. For instance, by us asking clarifying questions students realized the wide range of opinions about what made a teacher "good" and this highlighted the need for them to become more concrete about the characteristics they would want to measure. Now students were interested in counting things like how many students did a teacher send to the office, how many teachers were from backgrounds similar to themselves, and how many students would describe a teacher as "nice" or "respectful" or as someone who taught interesting classes.

The next step was to support students in developing their proportional reasoning, because this is one way that people mathematically measure bias. I introduced the concept of proportion by discussing the racial representations within a school's faculty. In our first full-class conversation about race, I discussed the racial composition of the teaching staff because students had already raised the underrepresentation of faculty of color at Champion as an issue of interest to them (with many students identifying teachers who reflect student demographics as a way to measure the strength of a school). I also hoped that because the topic was not *directly* connected to the students' identities, they would not feel targeted or unsafe compared to other possible topics. I presented students with two situations—one school with only three teachers of color, and one with 30—and asked them to

give their opinions about how that might make students in the schools feel, what messages it might send to the school communities, and how realistic those numbers might be. I then asked them to reexamine their responses if they found out, in the first scenario, there were six total faculty members, and in the second, there were 100. Lastly, students defined variables and built expressions that helped them measure what they found to be important with respect to the racial identities of the teaching staff at a school (ranging from the percentage of faculty who are persons of color, to percentages of faculty disaggregated by race and gender, to the proportion of faculty who shared a student's skin tone), and then they shared their work with classmates.

Because many students identified the importance of having fair school administrators and just discipline policies, their next task was to evaluate data on school disciplinary actions for different demographic groups. I entered this conversation cautiously since there had been a few recent suspensions at our school and, in informal settings, some students had already discussed amongst themselves how they saw a double standard based on race and gender in how discipline policies were applied.

As we examined state and national discipline data for different demographic groups, many students had the opportunity to see for the first time the numbers that reflected what they felt about the impact of disciplinary practices on different groups of students. For some, the apparent inequity they were witnessing and expressing in conversations was highlighted as they examined detention, suspension, and expulsion data for different racial and ethnic groups, socioeconomic levels, gender, and home languages. As students analyzed these data, it was imperative to have sound studies at hand that countered the reasons so frequently provided by the dominant culture—and often internalized by the rest of our society—for the disproportional representation of some groups in school disciplinary actions. So when Champion's students were confronted with the fact that "blacks [nationally] are now more than three times more likely than whites to be suspended"[1], it was essential to counter the common assumption that this was because black students are simply three times more likely to be disruptive at school. Instead, students discovered that there is no evidence of significant differences between African American and white students' behavior within a school environment, and that African American students "receive harsher levels of punishment for *less* serious behavior than other students [emphasis added]."[2]

The discussion that took place once we presented students with these data was invaluable, because for some the academic research findings confirmed what they had been raising in circles outside the dominant school culture, and for others it gave them insight into an alternative possibility that punishment is not always doled out fairly. It is our hope that using data and research to discuss these relevant and meaningful issues helped convince them of the power of mathematics to not only measure what was happening to people like themselves, but also to be a tool to help them ask and answer the questions behind their stories as well.

Of course, our classroom discussions did not always go as smoothly or result in such important insights. A particularly confusing conversation occurred when I asked students to examine data that showed higher rates of suspensions for students eligible for free or reduced-priced lunches than those not eligible. Many, if not all, of the young people at Champion were familiar with the free and reduced lunch program, yet rather than finding that this study showed that poorer students were disproportionately suspended, students in my first period class were nearly all outraged for the entire period that students were getting suspended for going to lunch, for choosing not to eat lunch, or for not liking the lunch provided by the school. No amount of reframing, redefining, or cajoling could shift their thinking away from the "lunch"

part of the demographic and into the meaning behind it. Therefore, to facilitate my students' understanding of the meaning behind the phrase "eligible for free and reduced lunch" we discussed why people (generally those in positions of power) choose to use this type of coded language as a way of referring to students on the lower end of the socioeconomic spectrum, with possible reasons ranging from not wanting to alienate people who live below the poverty line to avoiding explicitly naming the class and wealth divisions that are unjustly perpetuated in our society. In the end, my students decided to use labels that were clear to them and more honestly represented the data they were examining, and agreed to replace the terms "eligible" and "not eligible for free and reduced lunch" with "poor" and "not poor." I made this change in time for my third-period class where fortunately things went much closer to plan, and we had a richer and deeper discussion about the impact of class on suspension rates.

From here, the class continued examining real-world data related to educational inequities. The individual school, district, state, and national data connected to many items on their original lists and enabled them to practice proportions, define variables, and create expressions as ways to express and measure qualities like "good teachers," "fair leaders," and "no drama." Following our suggestions, students sought different ways to gather data to best measure what school qualities they defined as most important to determine a school's success. Students looked at data such as high-stakes test results, student surveys, "highly qualified" teacher designations, and college acceptance rates to measure "good teachers" and then created their own variables and expressions that measured the qualities they believed to be of greatest value. They reviewed ways to capture "fun" classrooms and began to articulate ways to measure technology access (number of computers per classroom, number of students per computer, and ways teachers and students use technology in classes), student engagement, graduation rates, and the range of course offerings.

Students became more thoughtful about how to quantify the "drama" in a school community and chose to examine the elements of a supportive school culture not only via student survey-type measures but also data related to discipline, attendance, how well the faculty reflects student demographics, and school size. And in each of these different cases, we encouraged them to think about how their measures might impact different groups within a school and find ways to evaluate those potentially different outcomes.

On the day the class agreed to host the end-of-unit webinar, virtually every student gave a PowerPoint presentation in front of class and discussed her or his five proposed measures of a school's success with us (me personally and virtually with Almanzia via Skype). Some students' presentations showed greater proficiency with proportional thinking than others, but all students identified measurable attributes of a school, defined variables that related to those attributes, and created meaningful expressions with those variables. Furthermore, students had become more critical thinkers about the world around them as evidenced by the school qualities they were now naming. For instance, one young man felt it was important to gather data on the age of students who drop out. Another student felt that it was important to know about the expulsion rate of students and to examine that rate for both students living in poverty and those who are not. A group measured suspension rates at a school for different genders as well as whether there was any connection between students arriving to school on time and their socioeconomic status, stemming from a concern that economic issues such as a family's child care situation, transportation challenges, and medical needs could impact students' arrival

times. Even the students who continued to list "good lunches" as a measure of a strong school eventually described how they would quantify that characteristic.

Their work showed they had gained a solid understanding of how to use variables, create and interpret expressions, and reason proportionally in a short amount of time and become more critical thinkers about the education system and our society at large. And while there are many qualities of a school that cannot be captured in numbers, our students had the opportunity to see how in this era of accountability and measurement there are in fact ways

---

### STUDENTS ARRIVING AT SCHOOL ON TIME

| $S_p$ | Students living in poverty going to school on time |
|---|---|
| $S_n$ | Students not living in poverty going to school on time |
| $T_p$ | Total students living in poverty |
| $T_n$ | Total students not living in poverty |

Percent of Students Living in Poverty Going to School on Time $= S_p / T_p \times 100$

Percent of Students Not Living in Poverty Going to School on Time $= S_n / T_n \times 100$

---

that quantitative data can be used as a tool to advocate for stronger and more equitable education for all students.

During a professional learning community, teachers were impressed by the quality of the students' PowerPoints, with one teacher remarking how challenging it was for his algebra students to work with subscripts on variables, while our students used them naturally to define different demographic groups. An end-of-semester, standards-based assessment also documented the improvement in students' algebraic thinking.

Although the unit supported students in gaining insight into how math can be used to investigate issues of educational equity and in developing language to help them name their lived experiences, it stopped short of having them actually collect and analyze data related to their measures. This choice to keep them from getting their hands "dirty" with the data prevented them from empirically discovering how their own school measured up, yet this decision continues to feel like the right one given the intended scope of the unit and the challenges of working with raw data. Furthermore, providing students with a space where they could use math to discuss and analyze injustices created an important opening for them to raise and take action on related issues in the future. For us, this unit provided insight into our students' powerful visions of strong schools and reinforced our belief that students can learn important math skills as they use them to make sense of the world. □

## ENDNOTES

1.  Losen, D. J., & Skiba, R. (2010, September). "Suspended Education." Retrieved August 2012, from Southern Poverty Law Center: www.splcenter.org/get-informed/publications/suspended-education

2.  Skiba, R. J., & Horner, R. H. (2011, Vol. 40, No. 1). "Race Is Not Neutral: A National Investigation into African American and Latino Disproportionality in School Discipline." Retrieved August 2012, from *School Psychology Review*: www.nasponline.org/publications/spr/40-1/spr-401Skiba.pdf.

CHAPTER NINETEEN

# *Radical Equations*

## BY DAVID LEVINE

*Radical Equations: Math Literacy and Civil Rights.* By Robert Moses and Charles E. Cobb Jr. (Boston: Beacon Press, 2001.) 233 pages. $21.00 hardcover.

In *Radical Equations: Math Literacy and Civil Rights,* veteran civil rights activist Robert Moses collaborates with journalist Charles E. Cobb to offer a stirring account of the Algebra Project, a reform initiative designed to help African American students achieve a high level of mathematical competency. The book raises important issues about both math education and the struggle for racial equity within our schools.

The Algebra Project focuses mainly on the middle school years, when Moses and his colleagues believe African American children must be prepared to enter high school math classes, which will open the door to higher education and technical careers requiring a strong math background. It encompasses new curricular materials, teacher training, the development of student leadership, and community involvement well beyond the scope of most educational reform efforts.

From a modest beginning in Cambridge, Mass., the program has grown into a national network with 18 sites, over 100 schools, and 40,000 students.

As a member of the Student Nonviolent Coordinating Committee (SNCC), Moses pioneered voter registration work in Mississippi during the early 1960s. Through his soft-spoken courage and patient encouragement of local leadership, he played a crucial role in building

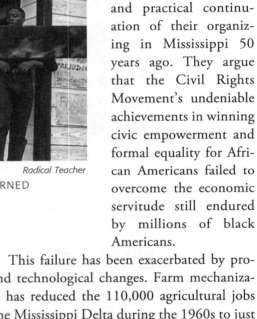

*Radical Teacher*

THE WORD "FREEDOM" APPEARS IN WHITE PAINT ON A CROSS BURNED OUTSIDE A FREEDOM SCHOOL IN PASCALOUGA, MISS.

the movement, which overturned state-sanctioned segregation and disenfranchisement in the South. After a sojourn in Tanzania, where he and his wife, Janet, taught school, he and his family moved to Boston.

By 1982, Moses had been tutoring his eldest child Maisha in math for years. He believed she was ready for algebra, a subject not offered for 8th graders at her Cambridge school. Since Maisha rebelled against having to do "two maths"—her regular schoolwork and the algebra tutorials her father insisted upon—he convinced her teacher to let him come to school to tutor her during the day. Soon he was working with a small group of students, and the Algebra Project was underway. As the program grew,

it also became a family collaboration—Moses' wife and children, Maisha, Omo, Taba, and Malaika, all came to play important roles. In the early 1990s, Moses convinced a colleague from his Mississippi days, Dave Dennis, to bring the program into the Delta. This work has grown into a multi-state "Southern Initiative" of the project, which Dennis directs.

In their book, Moses and Cobb (a SNCC field secretary in Mississippi from 1962 to 1967 and now a senior writer for allAfrica.com) present the Algebra Project as a spiritual descendant and practical continuation of their organizing in Mississippi 50 years ago. They argue that the Civil Rights Movement's undeniable achievements in winning civic empowerment and formal equality for African Americans failed to overcome the economic servitude still endured by millions of black Americans.

This failure has been exacerbated by profound technological changes. Farm mechanization has reduced the 110,000 agricultural jobs in the Mississippi Delta during the 1960s to just 17,000 jobs today, reflecting a national erosion in semi-skilled and unskilled jobs in the industrial sector. At the same time, the computer revolution has generated the need for "knowledge workers" with strong academic skills. Cobb and Moses contend that poor (and poorly educated) white, black, and Latino students of today are the equivalent of Mississippi's disenfranchised black sharecroppers of the 1960s, "trapped at the bottom with prisons as their plantations." More specifically, they argue that mastery of the increasingly technological workplace depends on increasingly sophisticated math skills, including algebra. "People who don't have it [algebra] are like the people who couldn't read and write in the industrial age," they argue.

To help African American students master mathematical literacy, the program has replaced traditional, rote-bound instruction with imaginative activities that engage student creativity and encourage sophisticated mathematical reasoning. An African drums curricular unit is designed to pair a drummer and a teacher in lessons which teach 4th and 5th graders about ratios, proportions, fractions, and rates. In his work with high school geometry classes, Moses encourages students to post their own versions of geometric proofs on the classroom wall, to be analyzed and possibly challenged by their classmates.

For the 6th-grade curriculum, which forms a bridge from arithmetic into algebraic thinking, Moses designed a five-step learning process. The students first observe or experience a physical event. For example, in a unit on positive and negative numbers, Cambridge students begin with a subway ride during which the teacher asks questions that focus their attention on their shifting environment. They then draw pictures, construct models, or in some other way create a representation of the event. The following step is to write a description of the event in their own language. Next, each class member translates her or his description into "regimented English," highly compact language which moves them into a mathematical mode, and from which they finally render the event as a mathematical expression. This five-step process helps students gain a firm grasp of mathematical ideas, connect math to everyday life, and become comfortable communicating in the language of mathematics.

Similar classroom practices in geometry and algebra courses encourage students to debate mathematical problems and actively construct their own understanding of math concepts.

The Algebra Project's pedagogy is not unique. It resonates with the experiential, inquiry-based approach advocated by the National Council of Teachers of Mathematics (NCTM) and resembles intellectually robust math instruction that can be found in some classrooms around the country. But the grassroots organizing philosophy of the program offers a dramatic departure from many mainstream reform efforts.

## The grassroots organizing philosophy of the Algebra Project offers a dramatic departure from many mainstream reform efforts.

Moses believes that math education innovations are often implemented by university researchers whose primary frame of reference is their own discipline and academic community, and the modus operandi is to offer prepackaged programs to schools. In contrast, the Algebra Project works on the premise that oppressed people can only win just schools through political organizing. To emphasize this perspective, the early part of the book describes how Moses and other civil rights activists built the Mississippi movement during the 1960s. With the guidance of Ella Baker, an experienced veteran of the black freedom struggle, Moses and his companions learned to develop the capacity of "ordinary people" to act as leaders and collaborate to bring about fundamental social change. Their approach, with its patient emphasis on democracy and nurturing the talents of poor people, has come to be known among civil rights historians as the "organizing tradition" of the movement. It is often contrasted with the "mobilizing tradition" of Martin Luther King Jr.

and other charismatic leaders, which is successful at turning out large numbers at demonstrations but often neglects the day-to-day work that builds powerful and sustained grassroots involvement.

For the Algebra Project, "organizing in the spirit of Ella" rests on three principles:

**1. The centrality of families to the work of organizing.** When Moses and other young organizers reached the Mississippi Delta, they connected with strong local leaders. Often, these leaders would involve family members in the movement, helping to create crucial networks of political activists. The Algebra Project seeks to involve the families of students and other community members in committees which run the local projects.

**2. Organizing in the context of the community in which one lives and works.** The young civil rights workers were absorbed into local families, who fed and housed them and protected them from hostile whites. This helped the activists "sink deep roots into the community." The Algebra Project also operates on the idea that staff members should be fully immersed in the communities which host local projects.

**3. Young people need to be empowered to fight for their own liberation.** Moses points out that high school- and college-age young people provided some of the crucial leadership of the Civil Rights Movement. He believes that the reforms necessary for young black people to achieve deep math literacy will only come about when they become ardent and savvy advocates for their own education. Through the program's Young People's Project, for example, students tutor their peers, lead workshops for students and adults, and help plan and run math youth camps during the summer.

## Algebra: A Key to Economic Liberation?

One of the book's core arguments is that students must master algebra to succeed in the workplace of the future. They cite Labor Department statistics that 70 percent of current jobs require "technology literacy" and that by 2010 all jobs would require "significant technical skills." Increasingly, essential technological expertise has come to mean relatively sophisticated understanding of how to use computers to perform a multitude of vocational tasks. To fully master computers, they argue, students need to be comfortable manipulating symbolic representations which represent "underlying mathematical concepts." They further argue that our society has designated algebra as the place where young people acquire such skills.

This cornerstone argument needs further documentation to be fully creditable. The phrases "technology literacy" and "significant technical skills" are quite general. We need to know if such literacy and skills specifically include algebraic thinking. Another issue is the varied impact of increased computer use on different occupations. The computerization of a job does not always bring the need for more sophisticated intellectual skills. Many low-paying service jobs have incorporated computer use which requires learning some new procedures, but not mastering substantially more demanding cognitive tasks. The authors would have been more persuasive if they had offered concrete examples of how algebraic skills are used in particular jobs, and evidence that such jobs are or will become a major part of our evolving economy.

Nevertheless, Cobb and Moses are not wrong to assert that algebra functions as a crucial gatekeeper to full economic opportunity. Even if a young person is not drawn toward a highly technical vocation, high school algebra is usually required for college entry. In addition, algebra provides knowledge necessary for advanced math which prepares students for a number of technical and scientific careers. Too many students of color lose these options through poor math performance before they reach high school. As Cobb and Moses note, part of this problem is reflected in Ph.D. statistics for technical fields. In 1995, blacks were 15 percent of the U.S. population but earned "only 1.8 percent of the Ph.D.s in computer science, 2.1 percent of those in engineering, 1.5 percent in the physical sciences, and 0.6 percent in mathematics." Finally, even though the authors could have presented stronger evidence regarding the relevance of algebra to adult employment, the technological evolution of many occupations does support their case. An understanding of algebraic concepts can help workers become more adept at working with spreadsheets, graphs, and databases. Our computer-based economy increasingly calls for such skills, even outside of highly technical fields.

## Is the Program Working?

In assessing initiatives such as Algebra Project, a crucial question is whether the program is meeting its stated goals.

In Bessemer, Ala., teachers at Hart Elementary, a school of mostly poor, black children, started participating in the Algebra Project in the fall of 1991 while teachers at the predominantly white West Hills, one of the "top elementary schools" in the district, continued with traditional math instruction. During a three-year study initiated in 1995, Hart moved from trailing West Hills on standardized math tests by several points to exceeding it by a few points, compiling the highest scores in the district.

*Radical Equations* and other Algebra Project reports are filled with similar success stories. They also document instances in which Algebra Project students register in greater numbers than their peers in higher-level math courses. As Cobb and Moses tell the Algebra Project story, they weave into their narrative extended testimonials from parents, teachers, and students which provide both penetrating explanations of the reform process and many examples of how the program has helped students learn more.

While the vignettes and overall narrative thread give us a persuasive picture of an effective reform movement, the book would have been

Staughton Lynd/State Historical Society of Wisconsin
A CLASS AT A MISSISSIPPI FREEDOM SCHOOL IN 1964.

strengthened by more systematic documentation and analysis of the program's impact on student achievement. We need to learn more about the extent of the program's success in strengthening students' math abilities, and the classroom dynamics which make such success possible. In-depth case histories of Algebra Project classrooms would be helpful, as would comparisons between the learning experience of students within the program and the learning experience of similar students in traditional math classes.

Research on the project should not fall victim to the popular and crude trend in American education to judge programs mostly by narrow quantitative measures. Cobb and Moses cite increased standardized test scores to document the program's success but realize that such numbers tell only a small part of the story. They examine the Algebra Project's impact on student motivation and work habits, teacher attitudes and behaviors, and community involvement. Future research should build upon and extend this holistic approach.

# An African drums unit pairs a drummer and a teacher in lessons about ratios, proportions, fractions, and rates.

## Algebra Project Lessons

In a review of school reform during the past century, educational historians Larry Cuban and David Tyack note that innovations often falter because their advocates fail to win political support. *Radical Equations* does a good job of teasing out insights from the kind of political work which builds durable support for substantive changes in how schools function. It offers a refreshing contrast to glib and self-congratulatory recipes for fixing up schools.

Even after nearly two decades of nurturing the program, Moses writes, "I have thought of the Algebra Project as a young child who is trying to stand up and teetering and falling down a little, then getting back up." The book pays careful attention to this teetering up and down of small groups of people trying to make their schools better. Cobb and Moses glean insights into the often contentious dynamics of school change from battles with the constraints of rigid standardized testing, uneasy administrators, and bureaucratic fear of innovation. The challenges faced by the Algebra Project affirm what they learned in Mississippi: People have to be willing to change themselves if they are to

develop the strengths they will need to change the system.

Beyond the issue of math instruction, the Algebra Project offers compelling lessons on how determined networks of educators, parents, and students can build a program which advances educational equity. Such democratic renewal promises the obvious rewards of promoting academic and vocational success for young people. But perhaps just as important, it also affirms local people's cultural values and capacity to deepen community life through shaping the public institution most likely to have a profound impact on their children. "Organizing in the spirit of Ella" means school reform which enriches the lives of teachers, community members, and students.

In contrast to top-down reform initiatives which demean the expertise and professional pride of teachers, Moses and his colleagues have developed training programs which build upon their strengths. A Cambridge teacher comments, "Bob was affirming what we were doing while he was helping us change. He didn't come in and say, 'We're throwing this out, it's junk.' He came in and said, 'You guys are great. Wanna try something different?' When we asked, 'How will it work?' he turned around and asked, 'Well, how do think it should work? What do you want to have happen?'" By posing problems rather than solutions, Moses invites teachers to confront and work through the frustration and anxiety of experimenting with new ways of teaching.

Such collaborative processes within the classroom are buttressed by efforts to involve community members. Although the dynamics of community involvement differ from site to site, the project is deeply committed to encouraging local control. During a 1998 visit to the Jackson, Miss., Algebra Project, I talked with Kathy Sykes, who served as a project staff member and representative on the local Site Planning Committee. This group reviewed the program budget, helped plan such activities as student retreats, and encouraged parents to serve as chaperones for program activities. The committee also encouraged parents to sit in on classes and eventually hoped to train parents as classroom assistants. Sykes told me, "I feel this is sort of like a crusade. ... I think that the work which is going on here will make a difference in the lives of our people and that's why I want to do what I can to see that it continues."

The program seeks to instill this spirit of personal responsibility through pedagogy which encourages students to break out of their own passivity and take charge of their own learning. Mary Lou Mehring recounts how 12-year-old student Andrea Harvey asserted, "I'm going to do four lessons a week because I want to finish such-and-such by the 7th grade, so that I can finish the book by the 8th grade, so I can be in honors geometry in the 9th grade." Andrea went on to work with the Algebra Project and eventually become certified to teach math in the Boston schools.

As a continuation of the Civil Rights Movement, the Algebra Project places itself firmly in the tradition of education aimed at racial equality. At the same time, Moses conceptualizes the goal of the endeavor almost exclusively as improved job opportunities. The program does not appear to directly use math instruction to help young people see full citizenship as the opportunity to use their math skills to promote social justice. As many articles in this book indicate, math can be used to analyze social inequities within our society—such topics as the disparities between rich and poor school districts, the mathematics of sweatshop economics, and the quantitative injustices built into the wealth and income structure of our society. Such themes might represent a fruitful direction as the program's curriculum evolves.

However, the absence of political math content hardly means the program is apolitical.

The authors persuasively echo Ella Baker's assertion that demanding something which is essential to your life and which you are systematically deprived of is an inherently radical act. Moses approvingly cites instances when young people agitate to make their schools dramatically improve math instruction.

For African Americans, the struggle for education has always been entwined within the struggle for freedom. This intimate historical relationship is underscored when the authors quote Mississippi school-desegregation activist Mae Bertha Carter: "The way to control black people or anybody is to keep them dumb. Back in slave time they catch you reading and they would whip you. Education, that's the goal. These [present day] school systems ain't doing nothing but handicapping these children."

In a society so afflicted with faulty historical memory, the Algebra Project demonstrates the necessity of learning from our past to fashion our future. In doing so, it puts history to its most honorable and practical use. □

## FOR MORE INFORMATION ON THE ALGEBRA PROJECT

**Algebra Project Inc.**
99 Bishop Richard Allen Drive
Cambridge, MA 02139
617-491-0200
www.algebra.org

**Young Peoples' Project Inc.**
99 Bishop Richard Allen Drive
Cambridge, MA 02139
617-354-8991
www.typp.org

# CHAPTER TWENTY

# Teaching Percent Change + Social Justice = Opportunity for Deep Mathematical Discussion

## BY FLANNERY DENNY

In my eighth year of teaching, I hit a wall with teaching percent change. Percent change is one of the few calculations taught in math classes that shows up regularly in the media, and one that I often do in my head to make sense of the world around me. Despite this, I had been teaching percent change using textbook problems about changes in the number of marbles in a jar, which had no meaning in my 7th-grade students' lives. I told myself that I could and must do better to engage my students' interest.

It was October 2010 and I had read that the representatives in the U.S. Congress were the most diverse that they had been in our nation's history and decided to see how dramatic the change had been over the course of my lifetime. I started with the 1979 Senate. How was it possible that nearly 60 years after women won the right to vote, only 3 percent of our senators were women?

And only one an African American? I want my students to believe that statistics can tell powerful stories about the injustices in our society, and these numbers seemed like a good place to start. I made a handout with the data in the box below.

I distributed the data to my students, asking for their observations and reflections. I hoped they would notice that our representatives in Congress do not represent the full diversity of the country, and question the system that is responsible for that divide. The room started buzzing.

"Are we learning math today?"

"Why is there only one black senator? That's racist!"

"Why are there so many more categories for the 111th Congress?"

"Why did you pick these Congresses?"

Each question was followed by reactions and hypotheses. I jumped in to engage with some, but I listened and let them try to answer

> ## "I want my students to believe that statistics can tell powerful stories about the injustices in our society."

each other's questions first. Some of their questions strayed far from mathematics, and we didn't arrive at satisfying answers to all of them. It's important not to get bogged down by this.

---

## STUDENT HANDOUT DATA

**95TH CONGRESS (1977–1979)**
<u>**Senate (100 members):**</u> 3 women, 1 African American

<u>**House of Representatives (435 members):**</u> 18 women, 17 African Americans

**105TH CONGRESS (1997–1999)**
<u>**Senate (100 members):**</u> 9 women, 1 African American

<u>**House of Representatives (435 members):**</u> 57 women, 41 African Americans

**111TH CONGRESS (2009–2011)**
<u>**Senate (100 members):**</u> 17 women, 1 African American

— 1 Hispanic, 3 Asian Americans, 1 Native American

— 13 Jews, 26 Catholics, 53 Protestant, 5 Mormons, 1 Eastern Orthodox Christian

<u>**House of Representatives (435 members):**</u> 78 women, 42 African Americans

— 25 Hispanic, 8 Asian American, 1 Native American

— 32 Jews, 161 Catholic, 239 Protestant, 2 Muslims, 2 Buddhists, 1 Quaker, 1 Atheist, 5 Mormons

— 3 Openly Gay

---

My students are a diverse group hailing from four New York City boroughs, vastly different socioeconomic means and a broad range of cultural and racial backgrounds. Although full tuition at the private school in which I teach is expensive, approximately 75 percent of our families pay fees calculated based on our sliding scale tuition model. I made a point of engaging the students' anger about the sole African American in the U.S. Senate. I asked them whether they think that people decide who to vote for based on race. They noticed race as part of the last two presidential elections, but they had a much longer list of reasons why people vote as they do. We agreed that race should not be the only criterion on which people base their votes, but we were clear that there should be more African Americans in Congress.

"Why do you think that African Americans have gained 26 seats in the House of Representatives over the course of my lifetime, but that there is still only one black senator?"

"There are fewer representatives in the Senate."

"You don't have to get the whole state to agree to vote someone into the House."

I reminded my students of the map of New York City's voting districts that we look at when we're lobbying our representatives. Keyshawn mentioned that most of the people in Harlem are black, so it would be likely that they would elect a black representative. Madeline pointed out that Chinatown probably elects an Asian American representative. I shared that the district I live in is predominantly Puerto Rican, and that our representative, Nydia Velázquez, was the first Puerto Rican ever to be elected to the U.S. Congress.

"What do you think happens outside of cities?" I asked the class. "Do you think that there are districts where most of the voters are African Americans?"

"Maybe in the South?"

I used the opportunity to explain that people in power have historically gone out of their way to make it hard for African Americans, Latina/os and other people of color to end up in control of voting districts by creating strangely shaped districts that disperse communities among several districts. The Voting Rights Act—the product of civil rights activism—attempts to counter that by requiring redistricting that gives each community a fairer chance. The data that we were looking at is a good reason why we have to be thoughtful about how we set up our voting districts.

It was time to move on, but I challenged my students to keep thinking about whether they could come up with a way to hold elections that would make it more likely for our representatives to reflect the full diversity of our society. Teaching social justice is never neat and tidy, and no teacher should anticipate arriving at the point where everyone's questions about the world we live in are answered. Life is complicated, and these questions are worth asking even if we can't always answer them.

## Calculating Percent Change

Next, I put the formula for percent change on the board and asked the class how we could use the formula to calculate the percent change of African Americans in the House from the 95th Congress to the 105th Congress.

$$\text{percent change} = \frac{amount\ of\ change}{original\ amount} \times 100$$

I encouraged them to think about what the words mean outside of math. We discussed what *original amount* means in this context and how we could calculate the *amount of change*. Once everyone felt comfortable with the example, I assigned each student a single calculation to complete and contribute to the chart on the board. I placed the elements on the chart so as to make comparisons easy.

The student assigned to calculate the percent change for men raised his hand and wanted to know how he was supposed to calculate when I didn't tell them how many men there were. I intentionally left if off the list; many students' thinking gets stuck in a box when faced with unfamiliar material. I stopped everyone and posed the question to the whole class. A student explained how to use the total number of senators and the number of women senators to find the number of men who are senators. The whole class relaxed because, even though it wasn't obvious to all of them, they all could have figured that out. I loved that they had to think a little and listen to one another to figure out what information to plug into the formula. In the real world, that's how math is. And it meant they were making sense of the formula and were more likely to remember it.

Looking at the completed chart (below) gave us lots to talk about.

We looked at the difference in the percent change experienced by men in the Senate vs. women in the Senate.

I asked: "Why is the same 14 seats turning over from men to women a 467 percent gain for women and only a 14 percent decrease for men?"

"The men started out with almost all of the seats, and they still have almost all of the seats."

"Women still don't have very many seats in the Senate, but they've gained almost five times as many as they started with. That's a big change!"

I pushed a little more: "Do you think that women feel differently now about their access to power than they did 30 years ago?"

"Yes!" a chorus shouted.

"What about men? Do you think they feel like they have less power?"

"No," they said, shaking their heads.

For me that is the essence of percent change. In all the years that I taught percent change via textbook problems, students never really understood why it mattered what you used as the original amount. I enjoy providing my students with a context in which to make sense of this.

"What do you think is a better way of talking about the progress that women have made in the Senate?" I asked. "They've gained 14 seats or their representation has increased by 467 percent?"

"Four hundred and sixty-seven percent sounds like a lot more change."

## PERCENT CHANGE IN CONGRESS

| **Women in the Senate**<br>Change from the 95th to 111th Congress<br>3 → 17<br>467 percent increase | **Women in the House**<br>Change from the 95th to 111th Congress<br>18 → 78<br>333 percent increase |
|---|---|
| **African Americans in the Senate**<br>Change from the 95th to 111th Congress<br>1 → 1<br>0 percent increase | **African Americans in the House**<br>Change from the 95th to 111th Congress<br>17 → 42<br>147 percent increase |
| **Women in the Senate**<br>Change from the 95th to 111th Congress<br>3 → 17<br>467 percent increase | **Men in the Senate**<br>Change from the 95th to 111th Congress<br>97 → 83<br>14 percent decrease |
| **Women in the Senate**<br>Change from the 95th to 105th Congress<br>3 → 9<br>200 percent increase | **Women in the Senate**<br>Change from the 105th to 111th Congress<br>9 → 17<br>89 percent increase |

"Fourteen seats doesn't sound like that much."

"I don't know, but women should have a lot more seats than we do."

"Maybe it's important to tell both."

# What do you think is a better way of talking about the progress that women have made in the Senate? They've gained 14 seats or their representation has increased by 467 percent?

There is so much more we could have talked about, but we were out of time. It was a lively class filled with deep mathematical discussions, and my students were thinking critically about social inequities. I was excited to have helped my students build more connections between math and "the real world." And I was confident that this lesson would contribute to conversations with peers at the lunch table, with the history teacher in current events class, and with parents at dinner.

## Choosing a "Real-Life" Question About Percent Change

But we weren't done. It is a priority for me to help my students learn how to connect their curiosity about the world to mathematics. Their homework assignment was to think of a question they would like to ask about the world that could be answered using percent change, and to find a data set with which to conduct the calculations. The goal was to make a poster about it during the next class. For some students this provided an opportunity to apply their new math skill to something that they were already wondering about. For others, this was a chance to think of a new question. Knowing that they

would be spending at least 40 minutes of class time engaging with their questions and that their work was destined for display pushed my students to be creative and to take the process of developing an interesting question seriously.

Students arrived at the next class with an exciting collection of questions and starts of questions. We began by sharing so I could give feedback on whether they were ready to move forward or needed to meet with me first. Hearing each other's ideas helped students focus on their own work by quelling their curiosity about their peers, and helped students who had struggled with the open-endedness of the assignment have a sense of the possibilities. The broad range of students' questions reflected the racial and socioeconomic diversity of the student body as well as their engagement with a broader curriculum focused on social and environmental justice, annual activism projects, and many weeks at the school farm.

The posters needed to include their question, the process of answering their question, the findings, an image, and the source of their information. The target time frame was to have posters ready to present for the following class period.

About half the class was ready to get straight to work and started gathering the materials they would use to make their posters. A few students had data sets but did not record the source of their data and were sent straight to the computers to find them.

The rest of the students needed my help. Some students asked to use the computers. Since we have only a few computers in the room, I wouldn't let students use them until I was convinced that they had developed their question to a point at which they could be efficient. While I conferenced with students one-on-one, I asked those who were waiting for me to work with a partner to improve each other's questions.

Kyree impressed me with a list of average baseball player salaries for every year since 1916 and I was confused as to why he felt stuck. It turned out that he had found so much data he didn't know what to plug into the formula. It hadn't occurred to him that he got to decide which years he was interested in comparing.

"That's the exciting part," I told him. "You're in charge and get to make decisions." I suggested that he pick a year from a long time ago and a recent year. When he decided to use data from 1980, I reminded him to compare data from this year with data from a long time ago.

"But Flannery," he objected, "1980 was 30 years ago!"

"You're right. It's fine," I said, feeling old for a second.

Since Daphne volunteers at the animal shelter, trick-or-treats for the Worldwide Fund for Nature, and focused her Bat Mitzvah on endangered species, it was no surprise that she wanted to make a poster about animals, but she didn't know where to start. I told her to key in to a single species and make a choice between telling a story of a species in rapid decline or of a story about an intervention that has made a difference. She liked the idea of a story of hope, so I mentioned a rehabilitation project with the California condor population that has been hugely successful. And she was off.

Keyshawn wanted to draw attention to factory farming but wasn't sure how to go about it. I suggested that there are many quantifiable components of government guidelines and industry definitions, and listed a couple.

He decided to focus on the space available to chickens.

Madeline was so excited about graduating that she wanted to do a project featuring her class. She needed to know how many people were in the first class to graduate from our school. I asked her where she thought she could find that information. She considered looking for a staff member who has been connected to the school since its founding in 1966, but then she remembered a plaque in the library with the names of the members of the first class and asked if she could go have a look at it.

For some students this was an opportunity to apply their new math skill to something they were already wondering about. For others, this was a chance to think of a brand-new question about the world.

Lucas wanted to focus on global warming but wasn't sure how to quantify it. He had a list of ideas, but didn't know which would be the best measure. I referred him to the 350.org project to see if he could understand why some scientists are looking at carbon dioxide parts per million as the best indicator of the health of our Earth's atmosphere.

Some students I conferenced with went off to find their data right away. Others needed more than one check-in before they met with success. One student was a few minutes late to his next class because he was recording data that he unearthed. But all students went home with the information they needed to complete their posters for homework.

During the next class period, students took turns presenting their posters to the group. They were as impressed as I was by the range of questions and shocked by many of the findings.

We looked at the percent change of:

- height of wind turbines since the 1970s
- salaries of professional athletes since 1980
- amount of sales of organic foods in the last 10 years
- number of men and women in the *Saturday Night Live* cast since the first season
- living space available to free-range chickens compared to factory-farmed chickens
- audience size at Metallica concerts in the last 20 years
- size of the California condor population since a rehabilitation program started breeding them in captivity in 1982
- assets of the Apple Corporation in the last year
- global population in students' lifetime
- number of undocumented immigrants deported from the United States since Obama was elected
- carbon dioxide in the atmosphere since the beginning of the industrial revolution
- the number of colors available in a Crayola crayon box since the original box of eight colors hit the market in 1903
- teen pregnancy rates since the 1990s
- size of the graduating class at our school from the first 8th-grade class to this year
- abortion rates since abortion was legalized
- number of Yunnan golden monkeys since the opening of the Baima Snow Mountain Nature Reserve
- number of mountains lost in Appalachia since the start of mountaintop removal coal mining

After school I hung the posters in the staircase. Over the next several weeks I saw students from younger grades stopping to discuss the project; some even sought me out to ask whether they will get to do this project when they are in the 7th grade. Teachers at the lunch table told me they were excited to see the work.

For my students, seeing their work on the wall elevated the importance of the skill that we learned. Although we spent only three 45-minute class periods on percent change, my students appreciated the diversity of its applications in the world. They surprised me by deciding to use percent change to make sense of data in other open-ended contexts, like math trail presentations or in preparation for the social justice data fair. Their interest in percent change will also make it much easier for us to learn about exponential functions in 8th-grade math.

Although this unit could be a stand-alone, for my students it does not exist in isolation; it is part of a series of ways in which I engage my students in making connections between social justice and math. And I am only one teacher in a community of educators at Manhattan Country School whose explicit mission is to create opportunities for our students to wrestle with themes of identity, community, diversity, sustainability, civil rights, activism, and justice. I hope that my students, as citizens of the world working to create positive change, will think of mathematics as one of the tools in their toolbox.

## ONLINE RESOURCES

Black Americans in Congress: http://history.house.gov/Exhibitions-and-Publications/BAIC/Black-Americans-in-Congress

Women in Congress: http://history.house.gov/Exhibition-and-Publications/WIC/Women-in-Congress

Hispanic Americans in Congress: www.loc.gov/rr/hispanic/congress/chron.html

Information on redistricting: www.publicmapping.org/what-is-redistricting/redistricting-criteria-the-voting-rights-act

---

Students' names have been changed.

# CHAPTER TWENTY-ONE

# The Square Root of a Fair Share

## School Desks and Dream Homes

### BY JANA DEAN

Man oh man, were they crowded. When I asked my 8th-grade class how much personal space they got at school, their eyes rolled and—no surprise—they started talking to each other. They felt cramped in the hall, leaving class, and pressing against the door in the cold in the morning. And pity the poor souls who get a middle locker on the bottom tier. In-class conditions weren't much better. At their desks, the bother was their stuff. Prepared students pack a binder, a planner, two pencils, a pen, a textbook, a notebook, and a book for silent reading. Students soon learn that part of getting along in class means taking up as little space as possible. Although 8th graders don't usually make the connection, how much room they get is mathematics: It's a matter of area.

After a few years of having my middle school students struggle to grasp square roots, I had decided to start with listening to how they experience area in order to connect squares and square roots to their everyday experience. I hoped that giving my students multiple opportunities to see that the side length of a square is the square root of its area, and conversely, that the area of a square is its side length squared, would help them understand this fascinating mathematical concept.

After my students' ready response to my query about personal space, I decided they would compare the area of their desks to the area taken up by their supplies. From there, I would invite them to think critically about how and where people experience luxurious amounts of space or pressing closeness every day. The class would compare the size of the average newly built suburban house to houses built a generation ago and to dwellings in other parts of the world.

Any time students made comparisons of area they made square scale representations. As unconventional as it was, using squares made it easy to both compare the area of dissimilar shapes and see side lengths of squares as square roots. It also provided a link to the convention of measuring area in square units.

**New Desks Anyone?**

The previous spring, our vice principal had come around and asked if anyone would like new desks. The offer lasted about 25 minutes and I missed it. I started my students on area with the story. "I didn't jump on the opportunity," I told them, "because I hadn't been well enough prepared to know whether the new desks would give you any more work space than you've already got. In case he comes around again, I want to be prepared and I need your help. You are the best people to figure out how much space 8th graders need on their desks."

I passed out rulers, and issued a challenge: "You have 10 minutes to measure and calculate

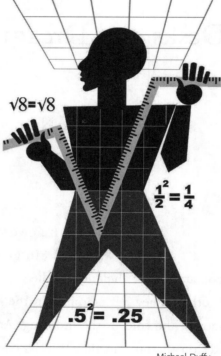

Michael Duffy

the total desktop area you need for your books and supplies."

As students finished measuring and began reporting their results to the class, Terra, with quiet drama, zipped up her binder to make the foundation of an elaborate tower on her desk. Ever prepared to learn, Terra always has several pencils, all sharp, and pens and highlighters, and she loves to read so she's got a reading book and a backup just in case. Once she'd finished her creation, she raised her hand with urgency. "Ms. Dean, Ms. Dean, look at all my stuff. There's no way I have space."

I turned to her decidedly undramatic desk partner Shane and asked, "Shane, ever feel crowded?" He nodded gravely.

It turned out that Terra needed more space than most: she required 640 square inches, while the class averaged 568 square inches.

**Seeing Squares and Roots**

In our school, students sit in pairs and share 8 square feet of desktop. That's 4 square feet each. To introduce modeling area with squares, I made a masking tape square on the floor with side lengths of 2 feet to represent the 4 square feet allotted to each student. After fielding questions before class started about the unusual square on the floor, I asked students to make a scale drawing of the square in their notebooks and use it to determine the number of square inches in a 2-foot by 2-foot square. Because my primary focus was the new concept that the side length of a square is the square root of its area rather than the previously learned concept of scale, I provided the scale. One unit on the graph paper was to equal one linear inch.

Once their squares were drawn, they could see that a sketch of a square with sides of 2 feet encloses 576 square inches. Some students laboriously counted squares, but most saw that 24 rows of 24 yielded 576 in total. As a class, students then compared the amount of space taken up by their supplies to the area of their desks and determined that they had on average 8 square inches of desktop elbow room to spare. Seeing how small that really was would wait until the next day.

I then asked them to select three integers between three and 10, and sketch squares using those side lengths. Most students quickly saw the relationship I wanted them to see, that is, the product of a number multiplied by itself can be represented by the area of a square.

Before school the next day, I taped off three 8-square-inch squares. I would use these little squares as a means to work back from squares to square roots: to find the side length of a square, you find the square root of its area. What's hard about square roots is that most numbers don't have a square root in the family of rational numbers, which includes fractions, decimals, and integers: We can only estimate the square root of 6, 7, or 8 unless we leave all or part of it under the radical sign. And leaving a number under a sign like that seems terribly unfinished: It just doesn't look like "an answer."

I put one of the 8-square-inch squares on Shane's desk, one on the floor at the front and one in the corner of the whiteboard at the back. They looked pretty small. During our discussion of the relative size of these squares, Shane tried fitting his elbow into the one on his desk. No luck. With his bulky sweatshirt, it was too small. His gesture elicited a chorus of, "That's not fair!" In order to build a bridge from their indignation to the math I wanted students to extract from their experience, I asked, "How do you think I was able to know how long the sides of those 8-square-inch squares needed to be so that we could visually compare 8 square inches to the 576 square inches of your desk?"

So far, I had kept silent on a little secret: the square root button. This little item on the calculator makes magic of finding square roots, and with its apparent ease keeps students from truly understanding a concept that at first is as new as division was when they'd never seen it before. Before handing out calculators, the most common mistakes surfaced, including dividing eight by two and by four. Two times four is indeed eight, but we needed the same number multiplied by itself to result in an 8-square-inch square. Someone suggested that since two was too small and four was too big, we ought to try three. No luck. A square with a side length of three has an area of nine. As I handed out calculators, I kept my carefully guarded secret, and privately thanked Texas Instruments for hiding the square root function in a second key. Five almost silent minutes followed as students used their calculators to try to find a number between two and three that would multiply by itself to get eight. Lexi got the closest with 2.83 squared coming to 8.0089.

Following this careful introduction, students spent about five class periods modeling squares on graph paper and finding the lengths of their sides. This exposed them to the difference between perfect squares like 9, 16, and 576, which have integers for the lengths of their sides, and squares such as 5, 8, and 17 whose side length can only be expressed exactly using a radical sign. Students became increasingly comfortable with both the radical sign and decimal approximations for side length. For this part of their learning I relied on our district's *Connected Mathematics* textbook, and a visual mathematics curriculum called *Math and the Mind's Eye* (see Resources on page 168).

## Room for the Stuff You're Going to Buy

By now, I knew my students had a good initial grasp of square roots, but I knew that asking them to apply their understanding in a new context would provide more practice for those who needed it and an opportunity to make interdisciplinary connections for everyone else.

I began setting the stage to analyze an area issue that I knew they had experienced right in their neighborhoods: the supersizing of new homes. The small cities surrounding Seattle have some of the highest population growth in the country. More new subdivisions punch into forests and farms every month. As though this pressure to house a lot of people weren't enough, it's amplified by an advertising-fueled drive to want more, which means house sizes have increased dramatically in the last few generations. Prior to the 2008 mortgage crisis, one of the largest regional builders, Quadrant Homes, boasted a home-building rate of seven houses per day and the slogan "More home for less money." On the internet, Quadrant Homes and others featured

Michael Duffy

prominent links to mortgage deals that seemed too good to be true. On the street, Saturday temp workers waved signs to the nearest show model. Since then, the building frenzy has slowed.

On a recent trip to Home Depot, which serves contractors as well as homeowners, I experienced a 5 to 1 salesperson to customer ratio in what had been a bustling, largely self-serve warehouse store. The supersized houses remain,

however, and new subdivisions were very much a part of my students' experience.

## It Hasn't Always Been This Way

I wanted to place the phenomenon of supersized single-family houses into an historical context. Until the 1970s, the average new home in the United States measured no more than 1,700 square feet. Just before and after the Second World War, they were even smaller. The 1940 average was 1,100 square feet. In contrast, new homes in 2005 averaged 2,400 square feet. And the size of the average household hasn't grown: It has become smaller as houses have grown. If you put the U.S. average household of 2.3 people into a brand-new 2,400 square foot house, everyone gets about 1,000 square feet. Compare that to the 1,100 square feet of the 1940s, typically shared four ways. Of course, averages don't necessarily reflect conditions on the ground. According to a quick in-class hand-raising poll, only about a fourth of my students lived in households with three or fewer people, with the majority living in households of four to seven. All the same, the national trend is more personal space.

In order to assess local home sizes, students tallied the square footage of the houses for sale in one day's classifieds. They ranged from as small as 1,200 square feet to as large as 5,900 square feet, with most of them falling between 1,600 and 2,400 square feet. Clearly homes in our town were bigger than the national average of two or three generations ago.

I asked students, "Why do you think houses have gotten so much bigger over time?"

Their responses were astute. Maya said, "It's that luxury thing people see on TV. They want status." Rickie echoed, "Yeah. They want to impress their friends." Alex said, "All that space is for the stuff." "What stuff?" I asked. "A couch? A bed? A dresser?" "No," he said. "Not what you already have. The room is for everything you're going to buy." Grant said the space was for the technology. "You know, we need room for our computers, our TVs, stereos, and everyone wants their own so we need separate rooms so we don't bother each other."

## My Big Dream Home

Great big brand-new houses weren't new to most of my students. Whether or not they'd ever been in one, they'd seen them. When I showed the class a picture of a housing development downloaded from the internet to get them thinking about what they see every day, Sean said, "I live there!"

In my haste to be ready for class, I'd downloaded the first image I found. It was one from the South. "What do you mean, Sean? This picture's from Georgia."

"No. I live there. That house is across the street from mine. There are seven of them just like that." He went on to describe a whole row of brand-new houses with huge brick entry facades. After Sean described the cul-de-sac facing his house, I asked, "Who do you think gains from these big new houses?"

Samantha snapped right to. "I think the builders of the houses gain because the bigger the house gets, the more it will cost, so the builder gets more money." Katy carried her thinking to who might lose from having such big houses. Like many of her classmates, she's seen woods consumed by development. She said, "Sometimes just two people will build a house big enough to fit 15 people. ... We knock down trees that give us fresh air ... like over by my house they probably knocked down 100 trees for more huge houses."

Soon Nate had his hand high in the air with an offer. "Ms. Dean, I draw houses all the time. Here, I've got plans right here in my binder." He pulled out detailed house plans for his dream home. It had two stories, complete with a library, media room, bonus room, sky bridge, breakfast nook, five bedrooms, and private walk-in closets and bathrooms for everyone. Nate hadn't drawn his plans to any formal scale, so there was no way for us to find the square footage, but all the same, it provided one representation of what the American dream home has become. After Nate shared his plans, I asked the class if they'd like to live in a house like the one he'd designed. Most of them thought it'd be pretty nice, especially the media room. Then I asked how many thought their current home was as big and fancy as Nate's dream. None did.

## Traveling the Material World

I wanted students to see that not only have house sizes increased in the United States, but they are also far larger than homes anywhere else in the world. Nowhere else on earth do people expect in such large numbers to have so much indoor living space.

Photographer Peter Menzel's *Material World* drives this point home with grace and compassion. The book consists of family portraits and statistics about life in 30 nations. Menzel created the book by moving in with each family for a week for photographs and interviews. Each visit culminated with the household moving all of its belongings outside the dwelling and posing for a portrait. When I mentioned his research method, "No way, I'd never do that!" rounded the room. When I asked why, my students said that apart from being too much work, it was too private. Accordingly, Menzel's photographs show a rare combination of pride and vulnerability on the part of his subjects. Even though the book dates from more than a decade ago, the power of the portraits endures.

I hated to do it, but I had to cut up the book, two books in fact. I found two used copies for less than $15 each. Working on our living room floor, my daughter and I mounted all the complete two-page portraits and statistics we could cull from the two dismantled books on construction paper and laminated them. Each set of statistics included the square footage of the family's dwelling and number of people living in it. This yielded 21 tabletop posters that students could examine in pairs.

The next day, I asked my students if they ever felt crowded at home. I planned to use their stories to tie our study as tightly as I could to students' own experience and to bring their attention to another layer of the mathematics: Comparing home sizes wouldn't have much meaning unless they considered the number of people sharing the space.

Lizzie told the story of a few years spent living in the city. She said, "Here in Washington we had a house. We moved to Oakland and we didn't even have room in the kitchen for a cutting board."

"Did you have the same number of people in the household?" I asked.

"Yes, and I had to share a room with my sister. We got rid of almost everything."

"So your living space was smaller and with the same number of people you had less space per person," I confirmed.

She nodded.

Marcos told about how his house used to have four people living in it, and now has eight. "That's half as much area for each of us," he said.

I then asked students to talk with one another about the relationship between the area of a home and the number of people living in it.

Zach summarized for the class, "If the area stays the same with more people, you have less space per person. Both the size of a home and the number of people in it decide how crowded it will be."

I then introduced the portraits from Menzel's book by proposing that looking at how people in other parts of the world live might help us understand what's happening with housing in our own country. After I passed out the posters, I directed students to record the square footage per person for each portrait they examined. Introducing square footage per person and making that the basis of comparison would keep students engaged in making meaning mathematically while engrossed in the social issues raised by the portraits. They were to come to me for a quick notebook check between each poster. This allowed me to assess their understanding of the math. I also got to have brief individual conversations with students about what they had observed.

At first, most of their observations had to do with the pictures. The 1994 photograph of Bosnia features a bombed apartment building and a father with a machine gun. Iraq proved popular, and perplexing to students. Instead of an image of war, Menzel's book featured a peaceful preinvasion extended family posed on the roof of a spacious home. The portraits gave rise to such interpretations as: "The Vietnamese family loves their house very much," and "In South Africa, a good home equals a good life," in spite of the cramped quarters in these places.

Gradually, the numbers began to tell a story as well. The South African family of seven shared 400 square feet. That's 57 square feet per person. In Vietnam, spacious by comparison, each person got 172 square feet. In a moment between checking notebooks, I sought out Lisa and shared with her another student's comment, "None of these people are rich. They are just like us." I asked her if she agreed.

"Well, yes in a way," she said thoughtfully. "Their circumstances have more to do with the country they're in than with themselves. But none of these houses are anywhere near as big as the U.S. average, so I also don't agree."

## Squaring the United States Against the Globe

After my class had spent a little more than two class periods examining the portraits and calculating the square footage of living space per person, I heard them discussing with each other the difference in size between the United States dream home and homes in the rest of the world.

I wanted to reinforce just how different that was, and I also wanted to bring students' observations into a discussion that could prompt some written reflection. On our last day with the posters, I announced that students would give presentations, and that in those presentations they'd convey the story told by the pictures and the story told by the numbers, and make a visual representation from which to draw some conclusions.

I told students to take notes in the form of squares representing the square footage per person, all drawn inside a scale square representation of the modern U.S. 2,400-square-foot house. Between presentations students made efficient use of the square root function on their calculators and their rulers to draw little squares inside their representation of an average U.S. new house. By the time they'd heard 15 presentations, and drawn in 15 squares to represent the living space allotted to 15 people living in other parts of the world, they still had lots of room to spare in their house. Before students left class that day, I asked them to write down one thought about what they had learned based on the picture in their notebooks. Two sentiments dominated: Americans are spoiled and lucky, and people in other places are poor and unlucky.

I felt disappointed. I wanted them to think more deeply, draw on their own experiences and those of their classmates, and include mathematical reasoning in their responses. The next day I asked students to respond in writing to some more pointed questions. I asked, "What do you think Americans give up in order to have big new houses?" and "Can everyone in the United States get a big house if they want

one?" I also asked explicitly that students use mathematics to support their reasoning.

Students commonly reached the conclusion that people trade time with their families in order to have more space, more money, and more stuff, and that having a small house could actually improve family life because people would be closer to one another.

Angel, who I knew had been homeless, wrote, "I'd never really thought about the size of our homes. ... I'd always wanted a big, big house ... but it doesn't benefit us to build such big

Michael Duffy

houses. They take up space we may need in the future. People may even go homeless because they can't afford one of the oversized houses. If 57 square feet per person is enough in South Africa, why can't we be satisfied with 500 square feet, which is almost 10 times as much?"

## Falling Short

Not as many of my 8th graders incorporated math into their written reflections as I'd have liked. It's possible I hadn't provided enough nuanced background from which to draw. Although comparing homes from around the world spurred thinking on the part of my students, I skirted the issue of unequal resource distribution within nations. A day spent on the range of living arrangements in the nations featured in Menzel's book would have taken advantage of students' enthusiasm for the portraits as well as paved the way for discussion of inequities in housing arrangements here at home. Big new houses, while ubiquitous, are available to only a small minority of people. Expanding this background might have supported more students to make math connections in their writing. It would also have integrated statistics concepts into students' learning.

I also tiptoed around having students describe their own housing arrangements. Their families' housing is something middle school students have no control over: It's purely a chance of birth. In the part of the county my school serves, housing ranges from small damp apartments and leaky trailers to luxurious gated communities. As one student said early on, "A house shows your wealth, just like clothes and shoes." On one of the last days of our study, I overheard Luke tease Ally for her enormous house "with servants," while Josh told me privately that his house was only 400 square feet, and he didn't want anyone to know. I would like to have supported my students in considering how our economic systems perpetuate the "good luck" and "bad luck" of social class in their own lives. The next time, I will do more to gently surface these issues from the outset.

Even with these shortcomings, teaching square roots by looking at living space connected a difficult topic in middle school math to real human concerns both inside and outside the classroom. On their exam, my students showed far better mastery of square roots than in past years and it opened their eyes to how the rest of the world might perceive Americans. As Ryan wrote, "It makes me sad because of how much we have compared to others." Caleb echoed, "They could resent that we're all rich, even though we're not."

For me, learning that students have only 8 square inches of desktop to spare helped me see the classroom from an 8th grader's point of view. I've also taken a cue from my young students and started celebrating the tight living quarters of my own family more than I complain about the mess. And I'll be ready when the next offer of surplus desks rolls by. □

## RESOURCES

Lappan, Glenda (2007). *Looking for Pythagoras: Connected Mathematics 2*, Boston: Pearson-Prentice Hall.

The Math Learning Center. *Math and the Mind's Eye*, Salem, OR. Available at www.mathlearningcenter.org.

Menzel, Peter (1994). *Material World: A Global Family Portrait*, San Francisco: Sierra Club Books. (A set of 12 posters made with images from the book along with a curriculum guide is also available from Social Studies School Service at www.socialstudies.com.)

Wilson, Alex and Jessica Boehland (2005). "Small Is Beautiful: U.S. House Size, Resource Use and the Environment," *Journal of Industrial Ecology* 9(1-2): 277-87.

# CHAPTER TWENTY-TWO

# The Geometry of Inequality

## BY ANDREW BRANTLINGER

I taught a summer course several years ago to a group of students in a neighborhood school on Chicago's north side. They were there because they had failed geometry during the regular school year. Most of the students were non-white, and 85 percent of the school's students were classified as low-income. Many of the students in my class were not on track to graduate in four years and had a history of failing mathematics.

When I interviewed my students, many of them said they didn't see how mathematics related to their present and future lives outside of schools. So I chose to teach a lesson I hoped would help them connect geometry to issues of fairness and see how math could be relevant.

I began the lesson by asking my students if they had heard of Rodney King and the riots that took place in South Central Los Angeles in 1992. The five African American students claimed that they knew who Rodney King was, while many of the other students did not. One of my female African American students explained that Rodney King "got beat by four white cops" and described a bit of what she knew about South Central. I added that the disturbances broke out in South

Central and other neighborhoods in Los Angeles after a mostly white jury acquitted these four white police officers.

After this brief introduction, I asked the students to work in groups to come up with what they felt to be average ratios of people to movie theaters, liquor stores, and community centers. Obviously, a community's resources comprise more than simply theaters and liquor stores, and might also include parks, quality schools, libraries, diverse shopping areas, and adequate public services like water and sewage, not to mention access to good jobs. But the figures I had came from a National Public Radio story that came out shortly after the riots in 1992, and I was really struck by the disparity in South Central between the number of theaters and community centers on the one hand and the number of liquor stores on the other. I decided to use these statistics as the core of my lesson.

Initially, three of the four groups were reluctant to come up with estimates for the three people-to-resource ratios. I think this was, in part, because I was asking them an unusual school mathematics question—one with no exact answer. A few students insisted that I had a "right" answer in mind, despite my reassurances to the contrary. In some sense they were correct; I *did* push them to justify estimates that seemed unreasonable to me.

## Movie Theaters and Liquor Stores

Next I asked the groups to use their people-to-resource estimates to make predictions about the number of movie theaters, community centers, and liquor stores in a "three-mile radius" in South Central circa 1992. Having used this activity with preservice teachers a few months earlier, I realized that estimates for the number of community places in South Central can be made without doing much mathematical thinking. I required my students to engage with the geometric issues that the problem presents. I hoped a better geometric understanding would improve students' critical understanding of the situation.

Because I also wanted my students to connect their everyday knowledge—of things (population distribution and size of city blocks)—to the problem, I presented each of the four groups with a map of Chicago and asked them to draw a circular region with a three-mile radius on it. Again, I wanted students to think mathematically, so I asked the class to figure out how many square miles and square city blocks would be in such a region.

Some students responded by drawing a square-mile grid on the map, while others used more ad hoc approaches to finding the area. As I went from group to group I heard them ask each other: How big is a square mile? Where is a square mile on the map? Can we use $A = \pi r2$ to find the area or do we have to count squares? When two of the groups asked me if $A = \pi r2$ would give them the same result as counting the number of square miles inside the circle, I responded, "Try to find it both ways and see if they give you the same answer."

Approximating the number of city blocks in a circular region with a three-mile radius fostered a good discussion, in part because there was some degree of confusion. That is, my students were used to thinking about linear blocks as a measure of distance but they were apparently not used to thinking of square blocks as a measure of area. One group of students asked me if they could find the area of a square mile by counting the linear blocks on the perimeter of a square.

A student in a second group initially argued that the area of South Central would be 48 blocks because the diameter of the circular region is six miles and there are eight city blocks per mile. But one of her groupmates disagreed, stating, "No, it should be 96 blocks because you have to fill the circle up on both sides of the diameter." I asked this second young woman to show us the 96 blocks on the map. Shortly after beginning her count, she looked up, smiled, and claimed that 96 blocks would be far too few to fill up the circular region.

When I returned to this group later on, they explained that they used π(24)2 to calculate the approximate number of blocks (1,810) in the circular region. This solution—using a radius of 24 blocks instead of three miles—was slick. I had not considered it myself. When I asked them if they were sure of their answer, they showed me that they could fill up a quarter of the circle with rectangles, which when combined would give an area of about 450 blocks, one fourth of 1,800.

Working through the geometric aspects of the activity took the better part of a class period. At the end of the day, I asked each student to jot down a new estimate of how many liquor stores, community centers, and movie theaters they thought were in South Central circa 1992. I was a bit frustrated that several students were reluctant to give estimates. I had assumed that this mathematical work would make answering this question easier. Four or five of my students insisted they would first have to count movie theaters, community centers, and liquor stores in their own neighborhoods in order to come up with an answer they felt comfortable with.

We didn't have time for students to count in their neighborhood, so I provided them with data from Evanston, a mixed-income suburb that borders Chicago to the north. When we revisited the South Central problem the next day, I put an overhead up that stated that Evanston was a "fairly typical community" with an area of 8.5 square miles with approximately 75,000 people, seven liquor stores, eight community centers, and three movie theaters. (I used the Evanston yellow pages to estimate the number.) The students were able to quickly figure out that approximately three Evanstons would fit into the area of South Central and that a little more than three Evanstons would have about nine or 10 movie theaters, 26 community centers, and 27 liquor stores.

As my students read these results to me I told them, "I see this activity as a critique of our society as a whole, and our government, and not of the people who live in South Central." I said, "According to National Public Radio, at the time of the riots, there were zero community centers and zero movie theaters." This surprised them.

Then I asked the class to guess how many liquor stores they thought there were in South Central when the L.A. riots took place. They called out a range of numbers from 5 to 100.

When I stated that the actual number of liquor stores was 640, Dele let out an "Ooooh …" and Maya exclaimed, practically shouting, "What!?"

José said, "Oh man, that's cool," and laughed.

## I wanted my students to connect their everyday knowledge of things to the problem at hand.

Dele said, "All they want them to do is drink." Then Dele added, "All they do is drink."

Upon hearing this second comment, I repeated that I did not intend for this activity to be a critique of the people who lived in South Central. "Look, there are no jobs," I claimed.

"That's why they be on the streets," responded Dele.

Later on, when students were writing up their individual responses to the activity, Dele's groupmate, Tony, asked me how I thought the overabundance of liquor stores in South Central was connected to the riots. I explained that I meant there was probably good reason to rise up, considering the lack of alternatives and the gross resource inequities between what people in South Central had compared to, say, people in Evanston. Tony looked me in the eyes and nodded, but I don't remember him responding to me verbally.

### Potential Pitfalls

When I reflect on my "South Central" lesson I see two pedagogical problems. First, I had not expected this lesson to potentially reinforce the dominant worldview that the problem with South Central is the people who live there. My students know that some white teachers buy into prevalent myths that blame working-class blacks, Latinos, and whites for being poor and other social ills. So, in hindsight, it makes a lot of sense that Tony asked me to re-explain where I stood on the root causes of the riots.

The next time I do this activity, I will bring in additional resources in order to clearly set up the sociopolitical context of the problem and to hopefully avoid the possibility of reinforcing negative stereotypes. Excerpts from an article by Mike Davis called "Who Killed Los Angeles?" that discusses the root causes of the riots— i.e., economic disinvestment, racist government policy—would be helpful in this regard.

Second, I shied away from opening up the political whole-class conversation that might have happened at the end of the lesson. The students reacted to the shocking numbers, but then I spoke *at* them about my beliefs. While I did make an attempt to open up the floor, I should have asked them to discuss what *they* thought were the root causes of the riots and what they thought about Dele's two conflicting comments that "All they want them to do is drink," and "All they do is drink." I also wish I had asked Dele who the "they" was he was referring to.

Instead of pushing my students to enter the conversation, I rushed on to the planned end of the activity in which I explained Brazilian educator Paulo Freire's notion of "reading and writing the world." And I ended the activity by asking the class to write individual responses to the (perhaps leading) questions: "Is this activity a meaningful way to begin to read the world using mathematics?" and "Does the use of mathematics help one understand more deeply aspects of our reality?"

In truth, the problem of opening up a political conversation was not entirely due to my lack of pedagogical expertise; three of my students made it clear to varying degrees that they expected me to teach the depoliticized form of mathematics they were used to. One of my African American students repeatedly opted out of critical activities like this one, choosing instead to work alone on geometry worksheets. I wondered if her actions meant she doubted my assertion that math could be socially relevant.

Despite its flaws, I believe this activity was powerful for many students. Many students did seem to think that we should discuss issues of fairness in school. Typical written student responses to the activity read, "We should have more jobs and more community centers, then you wouldn't have to worry about the riots," and "The government should put more money in [South Central]."

Finally, I think that Dele spoke for many of his fellow students when he told me they were learning "more about real facts" in critical activities like South Central than in decontextualized geometry activities. This intrigues me because I have heard many mathematics educators claim that traditional mathematics is as close to the truth or "real facts" as students will ever get. Clearly, Dele did not see it this way. □

---

The South Central activity was adapted from an activity developed by Eric (Rico) Gutstein. See page 173. All students' names have been changed.

# SOUTH CENTRAL LOS ANGELES

## RATIOS, DENSITY, AND GEOMETRY IN URBAN AREAS

BY ERIC (RICO) GUTSTEIN

This is a very open-ended activity (mathematically), and students may be disoriented; adjust as needed. It is a group project, meant to be given in parts, as shown. I suggest you finish by having students share their responses to part 4(a) with the whole class and discuss different views.

This can be done in grades 7 to 10; in later grades, you can make it more challenging by altering part 3a. It can take 2-3 periods, or even more. Students can use resources like maps, almanacs with populations, and rulers, if you want. Make sure students carefully explain their mathematical strategies, because there are no "correct" answers for these problems and many possible solution methods.

### PART 1

a) U.S. cities/towns have movie theaters, community centers, and liquor stores. If you were an urban planner and had to design and plan a small city or large town, what seems to be a *reasonable ratio* of people for each of these—that is, how many people for each movie theater (i.e., individual screens), how many for each community center, and how many for each liquor store makes sense to you? What are you basing your estimates on?

NOTE: The definition of "community center" is ambiguous. One suggestion is not to count churches, but include fieldhouses in parks, Boys and Girls Clubs, YMCAs, etc. For liquor stores, ask students to consider stores that depend on alcohol sales, unlike supermarkets that also sell wine and beer (and even liquor). But these definitions are open for negotiation. The main point is for students to explicitly state assumptions. Also, if students have trouble on ratios, ask something like: "Would you have one movie screen for each person? One for a million people?"

b) Think about an "average" city in the U.S. that it is relatively crowded (dense) with one- or two-family homes, as well as apartment buildings of different sizes. In your "average" city, *using your ratios from #1a above,* what would be a reasonable number of community centers, movie theaters, and liquor stores?

NOTE: For example, if a group's ratio for community centers is *1:5,000* and their imagined city has *175,000* people, then they'd need 35 community centers.

### PART 2

The year 2012 was the 20th anniversary of the Rodney King verdict* in South Central, Los Angeles. When the verdict was announced, and South Central broke out in rebellion/

rioting (depending on your perspective), National Public Radio reported on the actual number of movie theaters, community centers, and liquor stores in a "3-mile radius" centered at the intersection of Florence and Normandie avenues (the center of the rebellion) in South Central.

a) Using your work from Part 1, what would you estimate to be the number of movie theaters, community centers, and liquor stores in South Central?

NOTE: Students may draw on background knowledge and/or media stereotypes to answer. Even if they do, push them to use math to figure out how many people live in a geographical circle with three-mile radius in South Central, then use their ratios from part 1(a) to determine how many of each there should be. Maps, rulers, and data of students' own locales can be useful. The point is that students need to figure out about how many people live in that three-mile area in South Central, then use their ideal ratios to find what should be the number of each resource.

For example, in Chicago, students can figure out how many people live in an area of the same size in Chicago, given the city's size (using maps, not the web) and population (about 2.7 million). Students can then adjust to South Central, which is about 50% denser than Chicago. I've had students use rulers to measure maps of Chicago, compute an overall population density, then adjust for South Central.

## PART 3

As reported on the radio, there were 0 movie theaters, 0 community centers, and 640 liquor stores in the 3-mile radius area.

a) What is the *liquor store density* in the area?

NOTE: Here, density means how many liquor stores per person and/or per block. In South Central, 16 blocks is about a mile. A good question to ask is: If liquor stores are randomly distributed throughout the circular area, about how far does anyone need to walk from their house to reach a liquor store?

## PART 4—QUESTIONS

a) Ask students to write about what they learned. Some questions you could pose:

- Why do you think these data are the way they are?

- What math did you use? What math did you learn in this activity?

- How was the math important for you to understand the social and political issues?

- If this was your neighborhood, would you do anything about this situation? Why or why not? If so, what would you do? ■

* Background reading on the Rodney King beating and verdict available at www.rethinkingschools.org/math.

# CHAPTER TWENTY-THREE

# Plotting Inequality, Building Resistance

## BY ADAM RENNER,
## BRIDGET BREW, AND CRYSTAL PROCTOR

Media depictions of San Francisco show idyllic images of fog pouring under the Golden Gate Bridge or happy tourists riding cable cars, but rarely the mostly non-white neighborhoods of the east side. San Francisco public schools have a bad track record of mimicking this masquerade, with very low numbers of African American and Latina/o students making it to senior year, and less than a quarter of those who do, graduating with the credits to move on to college. Our high school, the June Jordan School for Equity (JJSE), is located on the east side of the city, and was started by a group of teachers and parents who were disturbed by the high numbers of black and brown youth being underserved and then dropping out. We are an intentionally small school with a focus on social justice.

Our commitment to send students of color to college means that they need a strong math education. As members of the math department, we believe, like Bob Moses, that math literacy in itself is a civil rights issue for students of color. We have seen too many "math

Michael Duffy

haters" end up in remedial classes in college, short-circuiting their career options.

The teachers who helped found our school were mostly from the humanities departments, and it is easier to imagine getting straight to a student's heart and experiences with a great piece of literature or history told from a non-oppressor perspective than it is to imagine the quadratic formula liberating anyone. Part of our school's mission is to help our students become agents of social change, so making explicit connections to social issues in math class is something that we try to do, though many math standards do not make this easy. Still, when the lesson involves important math skills, social justice, and something that will grab student attention, there is the potential that class will be exciting instead of mundane.

## The Scatter Plot Project

No one took making explicit social justice connections more seriously than Adam Renner, who started as a 9th-grade math teacher at JJSE in fall 2010 after many years as a teacher educator at the university level. In one of his first major projects, he had his students use math skills as a way to dig into a deeper understanding of the chasmic divide between rich and poor in our city. He wanted to shed light on the impact of economics and the structures of racism on education, housing, and job opportunities.

Adam began by introducing his students to Zip Skinny (www.zipskinny.com), a user-friendly website for finding and comparing data about local communities. Our students live primarily in three San Francisco ZIP codes: the Excelsior, Visitacion Valley, and Bayview/Hunters Point. Along with mining for data in these ZIP codes, Adam selected four other ZIP codes for comparison: the Mission (an eclectic, centrally located neighborhood), the Presidio (one of San Francisco's wealthiest neighborhoods), and the Outer Sunset and Outer Richmond (two neighborhoods along San Francisco's Pacific coast). He asked the students to record in a table the following data: median neighborhood income, percentage of high school completion or higher, percentage of bachelor's completion and higher, unemployment rate, and percentage of non-white residency.

The freshmen had to find these data independently using Zip Skinny. Then, in carefully constructed groups, they had to graph two different sets of data on the same coordinate plane in order to discover the relationship between the sets of data. One example of a scatter plot they created was comparing X = median income vs. Y = high school completion; another was X = college completion vs. Y = percentage of non-white residency. In this way, students could see what it means for two circumstances to be related or correlated, but not necessarily by cause and effect. They also saw the difference between a weak correlation (the points are spread out) and a strong correlation (the points are almost in a line), as well as the idea of positive correlation (one circumstance increases with the other) and negative correlation (one circumstance decreases as the other increases). As they were learning the mathematical terms for data analysis, they began to discover that math can describe and order their world.

## Seniors as Mentors

Crystal, who teaches Probability and Statistics, had a class of JJSE seniors who were completing a similar exercise using spreadsheets and the various graphing and analysis functions of Microsoft Excel. We decided to do a group activity with the seniors and the freshmen that would further develop basic math skills like plotting points as well as data analysis. We brought the three JJSE math teachers, the 60 or so freshmen, and the 12 prob/stats seniors together to engage in some cross-class mentoring and jointly discuss these issues.

The mathematical purpose for the seniors was to establish which data would pair well together and to be able to share that information with the freshmen. The mathematical purpose

## Our students began to explore what numbers reveal about the world they experience every day.

for the freshmen was to understand the spread of data in order to scale and label each axis and to plot points.

The social purpose for both age groups was to work together as a community in order to have conversations about the implications of the data. The data essentially reveal that people in San Francisco have different life experiences based on the neighborhood where they live, and that neighborhood is strongly correlated with race. We wanted our students to have conversations about the statistics that seem to prove the racism that many of them experience, what that means for their communities, and what they might do about it.

Crystal asked the seniors to prepare a short lesson—based on the data that both classes had discovered and analyzed—that would introduce a new variable and help the freshmen create a scatter plot. Her class prepared for teaching the freshmen by spending time in class talking and doing math before meeting with the younger students. She asked the seniors: "What makes a neighborhood?" The kids talked about the kinds of specifics that define the character of a neighborhood. For example, they mentioned the number of payday loan stores as well as the demographics (e.g., age, race, gender, education level). Each senior decided which variables to use and created their scatter plots themselves, both by hand and using Excel. This preparation allowed the seniors to feel comfortable with the math that they were about to teach. Crystal did not spend class time talking about how to interact with freshmen, but we will have that conversation when we do this project again in the future. We paired each senior with five or six 9th graders.

On the first day of the project we met in the cafeteria and the seniors led their groups in discussing the data and creating large scatter plot posters. Although the freshmen had been exposed to plotting points on a coordinate plane, the issue of how to scale each axis—"How might we label the X-axis for unemployment, which ranged from 2.1 to 5 percent, differently from how we would label it for median income, which ranged from $37,000 to $740,000?"—requires sophisticated logical reasoning skills that the seniors had to demonstrate.

Questions about what the data mean, not only how to plot it, led to some rich conversations. One group's data showed that neighborhoods with more women was correlated with less unemployment. 9th-grader Kari asked Shauna, her senior leader: "Isn't it more normal for women to stay home with kids while men go to work?" Shauna laughed gently: "Who goes to work where you live?" Kari was relying on a stereotype about gender roles that is simply not true in many households. Shauna had learned in her three years at JJSE how the increasing number of men of color in prison has led to neighborhoods where most families are headed by women who are the sole support of the household.

On the second day of the project we all met in the library so that groups could complete

their scatter plot posters and so that seniors could lead their groups in a discussion about the data. Some of this discussion centered on the math (e.g., how tight was the correlation and was it positive or negative), but the primary focus was conversations about what the data reveal about our city and what, if anything, we can do to shift some of these trends. During the presentations Mimi, a senior, talked about her group's scatter plot, which showed a negative correlation between non-white residents and resident stability (how long people stay in their homes). She said, "Gentrification might be a reason that more non-white people means more moving." In her small group Mimi discussed new condominiums that are being built in the Bayview district that are prohibitively expensive for many families who have lived in that area for multiple generations. Mimi had one of her freshmen speak after her. Although the freshman did not show a clear understanding of gentrification (yet), she was able to talk about plotting the points, which was a new skill for her.

The 9th graders had to consider what it meant when points were close together in a pattern, and what it meant when they were spread out. Sometimes, the data were confusing. There seemed to be a positive correlation between number of kids and wealth of a family, but upon further inspection the freshmen had not scaled the axes correctly, and what one may expect (that more kids means a family has less wealth, at least in most neighborhoods) turned out to be true.

The freshmen were noticeably rapt while being taught by the seniors in ways that they simply are not with us, their teachers. Suddenly the ability to plot and analyze data points became a tool for freshmen to engage with cooler upperclassmen. The seniors did a post-project reflection in their class, and one student wrote, "The freshmen listened to us because we are role models and they respect us, sometimes more than the adults." Another senior observed, "Students who would not participate during class with teachers and other classmates participated with us, which made teachers feel surprised yet excited at the same time." We were.

Bringing the freshmen into the conversation about the intersection of math and social justice was important, but the unintended result of the project was what it taught the seniors. The senior reflections showed that they felt like math scholars and community leaders during this process, even though as freshmen most of them were intimidated and uninterested in math. The seniors were moved by the sense of responsibility that comes with teaching. Marshall wrote: "I felt like a teacher because everyone was listening to me and everyone was looking up to me to understand how to plot points, how to read a graph, and how to compare categories."

### What Does It Mean? What Can We Do About It?

The math was straightforward, but the statistics about our neighborhoods were wrought with emotion. They revealed some upsetting truths about our city. In their final analyses, freshmen had the following to say: "The graphs show us that non-white residents of San Francisco have a difficult time with money and finishing high school and continuing to get their bachelor's degrees like white students." Sage's group worried: "It lets us know how the incomes are, and they are really low. It makes us feel bad because we should be getting a higher income and we have no idea what to do about it." Reactions ranged from, "We don't care about it," to, "The graphs show to us the need to start changing for the better." Regarding what to do, some students were fatalistic: "No matter what we do, it's not going to change." Others showed a spark of passion and indignation: "It makes you feel like there is racism in San Francisco. It makes us feel bad there are not equal rights. We should make our own city."

The statistics provided both the freshmen and seniors with a tool for discussion, but did not reveal something completely new. We have students who live in a housing project that frequently does not have running water; they turn on MTV to see people their age celebrating birthdays with celebrity entertainers and new cars. Many of our students' lives are characterized by violence—whether it be the gangs that operate in the neighborhood where our school resides, the complicated way some of our students must travel to school to avoid specific neighborhoods, and/or the cycle of violence

> We trust that our students, even the youngest, can learn about the real circumstances that surround them.

in several of our students' families. They enter high school knowing that their lives are different from the media's depiction of a U.S. teenager, and this math lesson just helps explain those differences.

To some it may seem irresponsible to show these statistics to 9th graders. But we trust that our students, even the youngest, can learn about the real circumstances that surround them. In fact, we feel strongly that it is dangerous *not* to have discussions with students about race and class. They have seen and lived with the injustices for their entire lives, but are programmed to believe that poverty and violence are natural or that members of their communities and families just make bad decisions that lead to these outcomes. By having conversations based on numbers we are giving them the analytical tools to decipher the deluge of these messages, which is a step toward them being able to change the world around them. Math offers us a chance to analyze our world.

This project gave seniors an opportunity to claim ownership of what they have learned about inequities during their time at JJSE.

When asked what we should do about these truths, one senior wrote that city officials should really understand these statistics: "It is possible to look at the ZIP codes on their own and act accordingly in each ZIP code, and not lump them all together. One solution for the whole city couldn't possibly help everyone." Another

them how to plot the(ir) inequality to build resistance. We are learning how to study in a way that puts students at the center of the academic experience. *They* are the curriculum. ☐

Names of students have been changed.

## The inequality our students experience is planned; it is plotted. And, so we, in turn, show them how to plot the(ir) inequality to build resistance.

senior vowed: "I want to go to college and earn a degree and come back and help these low-income communities throughout San Francisco to fight against this environmental racism."

As educators we believe that at least part of the solution is people coming together to learn. Students studying the circumstances of their neighborhoods is an activity in community building, and communities in solidarity may be the strongest antidote to some of these statistics.

The inequality our students experience is planned; it is plotted. And, so we, in turn, show

*This article would be incomplete without sharing the sad news that one of the authors, Adam Renner, died suddenly and unexpectedly in December 2010 at the age of 40. Before coming to teach math at June Jordan School for Equity in San Francisco, Adam was an education professor at Bellarmine University in Louisville, Ky. Adam was a true warrior in the fight for social justice and his loss has been devastating to many people and communities. Though the world was definitely a better place with Adam in it, it is in his honor that we will continue to fight.*

# CHAPTER TWENTY-FOUR

# Integrals and Equity

## A Math Lesson Prompts New Awareness for Prep School Students —and Their Teacher

### BY MEGAN STAPLES

An AP calculus class at a prestigious boarding school does not seem a likely venue for student reflection on privilege and wealth. But when I taught a group of academically inclined math enthusiasts several years ago, I made some important discoveries, and I believe I helped my students view their world with a wider lens.

It was my sixth year of teaching math and I loved it—the intellectual challenge, the complex puzzle that would suddenly sort itself when that key piece was put in place, and the interconnectedness that appeared among seemingly disparate ideas.

I took a new job at a small private boarding school in central Massachusetts. I was impressed with the way the school guided its 300 students to become moral, thoughtful adults. There was a schoolwide work program, so everyone pitched in to maintain the buildings and grounds. They had an extensive community-service program, where students volunteered at soup kitchens and nursing homes and periodically undertook larger projects such as building homes for Habitat for Humanity. The academics, of course, were rigorous and nearly 10 percent of that year's graduating class would find themselves with Harvard admission letters. The students were privileged, and the school knew it was preparing future leaders.

My AP class had a weekly "lab period," which provided a luxurious 90 minutes of class time compared to the standard 42 minutes. The class had this extra time to allow for AP exam test prep, but it also afforded the opportunity to conduct some more exploratory lessons, which I tried to do as often as possible.

As with my other lessons, I planned this one with objectives that were decidedly mathematical. We had been working with integrals, finding areas under various curves and calculating infinite sums, and I wanted to introduce my students to a real-world application of this mathematical concept by examining the GINI Index.

The GINI index is a measure of income distribution (or dispersion) across a population—in this instance, we considered households in the United States. In a single number ranging from 0 to 1, the index provides an indication of the equitability of this distribution. A GINI index of 0 corresponds to perfect economic equity—every household has the same amount of income. A GINI index of 1 corresponds to extreme inequity, where one household has all (100 percent) of the income and all others have none. These are theoretical extremes; in reality, the number lies somewhere in between.

The index is calculated by first dividing the population into income groups—for example, by quintiles (fifths) or percentiles (hundredths)—and finding the percent of total income "owned" by that group of households. The group that

represents the lowest quintile, for example, will hold less than a fifth of the total income (the accumulation of all households). The top quintile will have more than 20 percent of the total income.

At right is one graph of income distribution from 1996. The accumulated percentage of income is plotted on the vertical axis against the percent of households (horizontal axis). The point (20,4) means that the lowest 20 percent of households held 4 percent of the total income. The point (40,13) means that the lowest 40 percent (the two lowest quintiles together) of households held 13 percent of the total income.

The GINI index is twice the area between the 45 degree line (representing perfect equality) and the income curve. Thus it measures the "gap" between perfect equality and the state of affairs described by the income curve. Note that the curve always includes the points (0,0) to (1,1). Its exact shape is what changes, depending on the distribution of income.

## Estimating Income

After a brief introduction to the idea of income distribution, we began talking about income in the United States. Perhaps not surprisingly, my students made reasonable estimates when I asked them to guess what percent of total income the top 20 percent of the population held. And they made somewhat reasonable estimates for the other quintiles. They didn't seem troubled by the fact that the top 20 percent of households held 47 percent of the wealth. (I assumed they attributed the disparity to super-rich individuals like Bill Gates and Michael Jordan.) But the most meaningful conversation arose when I asked them to consider the median income of the various quintiles. The median represents the middle value of the set of incomes in the quintile, often close to the average. I was perplexed by their lack of reaction to the percentage distribution, and I was curious what they thought the median incomes might be.

"Median income of the lowest quintile—what do you think?" I asked.

"Eighteen thousand," someone guessed.

"Twenty-two thousand," another said.

I recorded their estimates on the board. I didn't ask for any justification, leaving their reasoning private. Looking back, this is somewhat surprising as I was always asking my students, "Why?" and, "How do you know?"

I then asked them to guess the median income of the top quintile.

"One-hundred seventy-five thousand."

"Yeah, around there."

"One sixty." "Two hundred."

And others offered thoughts.

"And how about the next quintile down—this 60 to 80 percent group?" And we went through the rest. Once we had their set of values on the board, it was my turn to tell.

"Median income of the lowest quintile is $8,600," I reported.

The response was, "Whoa," followed by some silence.

"That means half the households are below that—10 percent of the people." Although this fact seemed to be mind-blowing for many, they did not challenge it.

I continued. "The next quintile up is $21,097. The third quintile is $35,486. The fourth quintile is actually $54,922, and the top quintile is $115,514."

My students expressed disbelief. "No way!" "Is that right?" I think the median incomes of the fourth and fifth quintile surprised them the most: They had thought that the fourth quintile would be around $90,000 and the top quintile would have a median income closer to $170,000.

Given their privileged backgrounds, perhaps I should not have been shocked that the students' conjectures were so far off. And perhaps

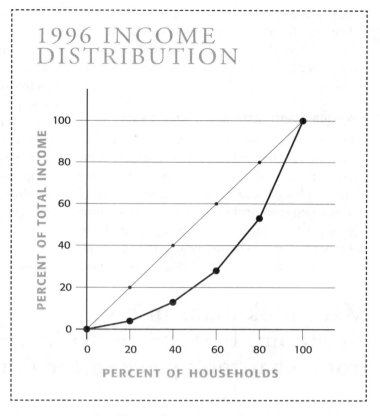

## 1996 INCOME DISTRIBUTION

I should not have been taken aback at their disbelief upon learning the median incomes for each quintile. Though I don't know the actual incomes of these students' families, this group was firmly situated in the top quintile, and most the top decile. These data were news to them—top headlines.

But I was surprised, perhaps most surprised, by the fact that the students had so little sense of their economic positions relative to the rest of society—this was despite all their community service, attention to issues of race and class in their history and literature classes, and interactions with classmates who were there on scholarships. Even with these experiences, the students inhabited a world where they could think of their families as more or less "average." In their social context there was very little evidence that would have led them to believe that only one in 10 people was as economically advantaged as they were. I don't know that this group of students had ever been asked to look at the "cold, hard numbers" before and pinpoint their positions relative to others. They knew they were from privileged backgrounds and had "more," but this quantification of how much more they had—orders of magnitude—was quite a surprise.

## Worlds Apart: Math and Social Justice

The experience of seeing these students gain awareness of their positions in society helped me along in my evolution as a teacher. I had gone into teaching because I loved math and loved supporting students in their journeys to become good people living in a challenging world. I also wanted students to be attentive to their communities and others around them, to understand

> Math took place in my classroom. It existed a world apart from what took place in the dorm.

that the world does not offer all an equal playing field. Ideally, I hoped they would use these understandings to change the world in ways that helped address some of the gross disparities that exist. But these were separate aspects of my "teaching." Math took place in my classroom. It existed a world apart from what took place in the dorm, on the soccer field or basketball court, during the van ride to the soup kitchen, or over an ice cream at the favorite local shop.

In retrospect, it's odd that these were separate worlds for me. I had been a math major at a progressive and politically active university. I had participated extensively in volunteer activities and lived in the Public Service House, a dormitory whose members were committed to community service. But math and being a good person in the world were separate for me there as well. My aspirations for social justice had not had a place in the math classrooms at the university, and I didn't bring them with me to the mathematical part of my teaching. These worlds

were not in opposition, but I didn't see them as natural bedfellows. Math was about the intellect and individual power. Being a teacher for me was about giving to society and connecting with individuals in ways that might make transformation possible.

It became clear to me that there was a large gap here, for me and my students. Despite my prep school students' general awareness and participation in activities, they remained unaware of their positions in that landscape. The hard and fast numbers brought this into relief.

I would like to say that I followed up on this initial lesson right away and implored students to examine other data that helped them understand themselves anew. We could have pursued the difference between men's and women's incomes or the incomes of different racial groups. We could have compared the United States to other countries. We could have looked at historical trends in the GINI index, and analyzed the factors that contribute to the greatest and least equity in income distribution over time. Regrettably, I did not pursue any of these.

Nonetheless, I was left with a powerful lesson. Through this activity I realized that those parallel pursuits could serve one another—where mathematical learning goals, even on a high level such as AP calculus, could be pursued at the same time as goals aimed at personal growth and social awareness. □

---

For another lesson on the GINI coefficient that can be used in calculus class, see: http://courses.ncssm.edu/math/apcalcprojects/econ/Gini_Index_Student_Handout.pdf

For further info and updated GINI coefficient data, including the multiple ways of computing it, and for different countries, see: http://en.wikipedia.org/wiki/List_of_countries_by_income_equality and http://inequality.org/unequal-americas-income-distribution

# HIV/AIDS STATUS

## USING HIV/AIDS STATISTICS TO UNDERSTAND RATIOS

BY SAGE FORBES-GRAY

Teacher Aim: Students will be able to interpret a statistic about HIV/AIDS transmission using their understanding of identifying, converting, and comparing ratios.

Student Aim: How do we understand HIV/AIDS statistics using ratios?

### PROBLEM OF THE DAY

a) 1 in 250          b) 1 in 500

What are two different ways of expressing each of the ratios (a) and (b)? Which ratio is greater? How much greater?

The whole class will go over the problem of the day, focusing on equivalent ratios, converting ratios, and comparing ratios. Possible answers for (a) will include 2 in 500, 1/250, and perhaps even 0.004 or 0.4 percent, and for (b) 2 in 1000, 1/500 and perhaps even 0.002 and 0.2 percent.

Hopefully students will see that 1 in 250 is greater than 1 in 500, and ideally students will see that 1 in 250 is twice as big as 1 in 500. Conclude by asking students: Where do they think these numbers come from?

### MINI-LESSON

Students will be told that these ratios have to do with HIV/AIDS transmission. Be sure they understand the language we will be using about HIV/AIDS. If there is time, we will brainstorm briefly about what we all know about HIV/AIDS. We will use a "set-up" that lists truth, questions, and myths (one column for each).

Ratios can be used to describe statistics and are used to describe statistics involving HIV/AIDS. The aim will now be elicited.

The following slogan was plastered on the sides of New York City buses as part of a public health campaign called "Know Your Status":

"1 in 250 people in the United States have HIV, 1 in 500 know it."

I tell the students that we will be answering a few questions about a statement that I am about to put on the board. I will ask the students to copy the following questions and tell them that they will be working in pairs (or groups of four, depending on the maturity of the particular class).

### QUESTIONS

1. What does this statement mean?

2. How could we restate this statement in three different ways? Try to use your understanding of equivalent ratios and converses to write these statements.

3. Why do you think many people find this statement alarming?

4. What can be done to change how people think and act about HIV/AIDS that would make this statement untrue, for better or for worse?

5. What kind of statement would be a less scary and more hopeful statement? (You do not need to use the same numbers for this statement.)

6. Do you think that this statement, which appears on the sides of many New York City buses, is a good way to get people thinking about the dangers of HIV/AIDS and the importance of knowing your HIV status? If not, how could the statement be

improved so that more people understand and are thinking about HIV/AIDS transmission?

As a class, we would review each question, focusing on any difficult language and reviewing equivalence and converses.

### CLASSWORK

The students will be put in their groups and given 15 minutes to discuss the questions and then 15 minutes to write up their responses. I will tell the students that at the end of class we will take 15 minutes to discuss what each group thinks about the activity and that they should be prepared to present for two minutes and answer questions. The presentations will be informal: Students will simply share their answers to one or two of the six questions.

Next, I would reveal the statistics, put the statement from the bus ads on the board, and begin to move about the room answering questions and making sure that the students are on task. I would remind them about their classwork grade and presentation at the end of class.

### FINAL PRESENTATIONS

I would encourage students to make connections between this material in their own lives. Finally, I would explain the homework assignment, offering a sample and answering any questions. I would collect the work as they get ready to move on to their next class.

### HOMEWORK

The students will be asked to write a letter to the campaign manager who designed the subway poster that they discussed in class today. In their letter, they would answer question 6 from the classwork and give their opinion to the campaign manager about how the subway poster affected them personally and how they think kids their age would be affected by the poster. The letters will be mailed to the campaign as a way to give the campaign manager real feedback from NYC youth. ■

## HIV STATUS

### BUS-STOP PHOTO TEACHING SUGGESTIONS

Have students research the number of people who are aware that they are HIV-positive. Then have them use the statistic on this bus-stop poster from the U.N. to calculate how many of those people don't know their HIV status.

Ask students to compare the statement in this photo with the ratio from the advertisement quoted in the activity above. Have them discuss why there might be a difference in the two statements.

# JUSTICE FOR JANITORS

## RICH LESSONS IN THE POWER OF MATH

BY BOB PETERSON

The picture book *¡Sí Se Puede!/Yes, We Can!* by Diana Cohn is a great picture book to use with students to show that math matters. The bilingual English/Spanish book, beautifully illustrated by Francisco Delgado, tells the story of a janitor strike in Los Angeles in 2000, when nearly 8,500 janitors put down their mops and brooms and demanded a living wage and dignity. The solidarity of many people helped the janitors win their three-week strike, which in turn sparked organizing drives in cities across the nation. As of early 2005, more than 200,000 janitors, most of them people of color, were organized into the Service Employees International Union (SEIU).

THIS ILLUSTRATION FROM *¡SÍ SE PUEDE!* HIGHLIGHTS THE POWER OF WORKING PEOPLE.

The text of the book tells the story of the strike from the perspective of the son of a single mother who is a janitor. While the book itself does not include much explicit math, it provides a context for rich lessons in the power of math.

### DO THE MATH

Students can work on understanding large numbers, multi-digit multiplication, problem-solving, and percentages through examining data such as:

• Because of the strike, 8,500 janitors won a pay raise from $7.20 to $9.10 an hour. What percent increase was this? How does this compare to the rate of inflation?

• Assuming a 30-hour work week, how much money did a janitor earn (before taxes) with the old wage and with the new one? What is the difference?

• Many of the janitors are forced to work only part time and thus don't earn enough money to prevent their families from living in poverty. In 2012 the U.S. government said a family of four had to earn more than $23,050 per year to be considered above the poverty line. Even on the new salary of $9.10 an hour, if a person with a family of four was only able to work 30 hours a week, how much below the 2012 federal poverty threshold would the family be?

• In the book the mother says to her son: "Even though I work full time as a janitor, I also have to clean houses and wash clothes on the weekend. That means we don't have any time together. And I can't afford to buy the medicine Abuelita needs to help her sore bones feel better." According to the SEIU, most janitors get no healthcare benefits from their jobs. In 2012, the average cost of a family health-insurance premium was $15,745 a year. How does that figure compare to the salary of a janitor working at $9.10 an hour for 40 hours a week? ■

---

For additional information on Justice for Janitors, search under "Justice for Janitors" on the SEIU website: www.seiu.org/a/search.php. For additional teaching ideas using this book, see www.rethinkingschools.org/math.

# CHAPTER TWENTY-FIVE

# Math, Maps, and Misrepresentation

## BY ERIC (RICO) GUTSTEIN

What happens when students begin to question what they have been taught in school? What other questions does this help them raise, and how does it help them to better understand the world? How can teachers create these kinds of experiences?

I considered such questions as I taught my 8th-grade mathematics class at Rivera, a public school in a Mexican immigrant community in Chicago. As part of my job as a faculty member at the University of Illinois–Chicago, I taught one middle school math class at Rivera on a daily basis for a number of years. My goal as a teacher has been influenced by the work of Paulo Freire and other critical educators. I want to help my students learn to read (i.e., understand) the world—through learning and using mathematics—as a way for them to begin to write (i.e., change) the world.

## Going Beyond Mathematics

As the basis of my curriculum, I used *Mathematics in Context* (MiC), an innovative mathematics curriculum developed in accordance with the National Council of Teachers of Mathematics standards. I found that MiC helped students learn to use mathematics to understand the world, as it developed many critical mathematical reasoning skills, but by itself, it was not enough. Therefore, I developed 17 real-world mathematics projects that my 7th- and 8th-grade students completed over the almost two years I was their teacher. (I moved up to 8th grade with my class.) This story, from the spring of 1999 when they were in 8th grade, is about one of those projects.

One significant lesson I learned is that going beyond mathematics is important in helping middle school students read the world with mathematics. Teachers need to develop a classroom culture that incorporates reading the world and examining injustice and oppression. An important part of going beyond mathematics is to try to normalize politically taboo topics. For example, my students and I had many conversations about race and racism, and they were central to a number of our classroom projects. I found that such an orientation is vital for students to appreciate and be more interested in mathematics, because they begin to see that mathematics can help them make sense out of their surroundings.

## Analyzing Map Projections

The project I describe here was called Analyzing Map Projections—What Do They Really Show? Maps are two-dimensional representations of the earth that we often take for granted. Few of us think that our standard maps might be woefully inaccurate, and we do not often consider how the images students see everyday on classroom walls shape their perceptions of the world. Mathematics is central to mapmaking, and different mathematical ways of representing the world produce very distinct maps. A goal of the map project was for students to use mathematics to analyze diverse map projections and to raise questions about what the various maps showed—and why. A larger goal of this project was to help students develop a more critical outlook towards knowledge in general.

I used two very different projections: the Mercator projection (developed in 1569 in Germany), which is the traditional map in U.S. schools (including Rivera); and the Peters projection (developed in 1974 by Arno Peters). The sizes, shapes of land masses, and coloring schemes of the maps are quite distinct. The Mercator map was developed during European expansion when colonial exploitation required that maps be used to navigate accurately (so as not to repeat Columbus' blunder) and was used successfully to find new territories. All maps unfortunately are misleading because they are two-dimensional projections of a sphere—and the Mercator suffers from serious visual distortion by altering the relative size of land masses. This is because the scale changes as you move away from the equator. Thus countries far from the equator (e.g., Greenland) appear much larger than they are. (Some Mercator maps have, in fine print, an explanation of this distortion and the mathematical information necessary to find the actual areas.)

For example, Mexico is about 760,000 square miles, Alaska is about 590,000, but Alaska looks two to three times larger because it is farther from the equator. The representations of Greenland and Africa are more distorted. Greenland, at 840,000 square miles, appears roughly comparable to Africa, which, at 11,700,000 square miles, is about 14 times larger. In addition, Germany is near the center of the map, which may have made sense from the perspective of European expansionism. However, since Germany is in the northern quadrant of the earth (Berlin is 52 degrees north), the only way to make it the center is to push the equator approximately two thirds of the way down the map. This compresses the Southern Hemisphere and enlarges the Northern.

# TERRITORY THAT MEXICO LOST TO THE UNITED STATES

## MAP TEACHING SUGGESTIONS

Using this map as a guide, students can use atlases and internet sources to measure and tabulate the total square miles of land that the United States took from Mexico following the war of 1848.

### DO THE MATH

Have students calculate the percentage of Mexico that was taken and the percentage of the continental United States that was originally part of Mexico.

For the land taken, the United States paid $15 million. Students can calculate the price per acre.

### VARIATION

Another option is to give the students no data other than Wyoming's area (about 100,000 square miles) and tell them to use that as a base for finding out how much land the United States took from Mexico. For deeper study, students can refer to the chapter on this war in Howard Zinn's *A People's History of the United States.*

These distortions remain true of today's Mercator maps, even though we now have better navigational means than the Mercator to sail across the Atlantic. Whether or not Mercator meant consciously to diminish the South, it is worthwhile to focus on the effect of the map's widespread use today.

In contrast, the Peters projection was developed to fairly and accurately portray the earth. As Ward Kaiser writes in an explanation of the Peters projection, "Peters is ... clearly focused on justice for all peoples, recognizing the values and contributions that all nations and all cultures can bring to the emerging world civilization."

The Peters map distorts shapes somewhat, but unlike the Mercator, it accurately presents the relative size of land masses.

As part of our project, I gave each group of three students a large Mercator map, borrowed from classrooms in the building. I gave them Mexico's size to use as a unit of measure and turned them loose to measure and compare several areas on their Mercator map (for example, Greenland and Africa; Mexico and Alaska). They used a variety of mathematical means to estimate the areas. Some traced Mexico and estimated how many Mexicos fit into other countries, then they multiplied the area of Mexico by the number of Mexicos that they estimated would fit into the other countries. Others overlaid centimeter grids on top of Mexico to determine the area per square centimeter, then measured other countries and found their areas. And some students used a combination of methods, including reallocating areas to make rectangular shapes whose sizes were easier to estimate. Afterwards, they dug through our almanacs to find the real areas of the places they had measured and compared those to their estimates. Finally, I had students respond in writing to several questions, both in their groups and individually.

## Questioning Past Learning

My students determined for themselves that the Mercator map did not show equal areas equally. This was further confirmed when they examined the Peters map. They were astonished and upset to discover the Mercator's visual distortions, and the majority felt they had been "lied to." Many were quite concerned about the implications of this misinformation, and some questioned other sources of knowledge and things they had learned.

In particular, many wanted to know why they had been mistaught. As one of my students, Rosa, commented, "The questions raised in my mind are why teachers never told us how wrong the map was."

Another student, Lupe, took a position of advocacy: "I think it's sad that we've all been taught this way. We should make our analysis public and let it be known. I just want to understand what [is] the point exactly of Mercator's map. What did he want us to believe, to see?"

One of the questions I had asked was, "Knowing we were all raised on the Mercator map, how does that make you feel?" In response, Marisol commented that it "makes me think

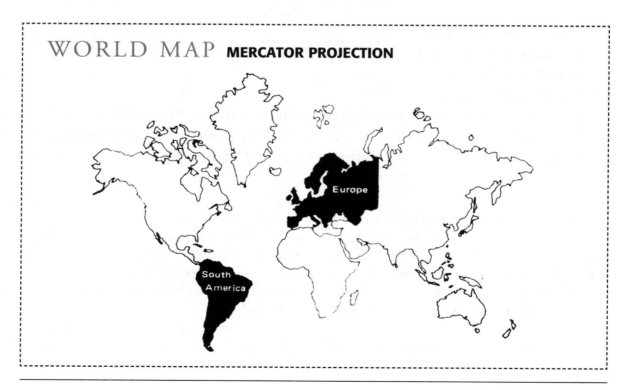

WORLD MAP **MERCATOR PROJECTION**

what other wrong information we have been given since childhood. It makes you doubt your social studies book, history written by the white people." And Gloria summed up the feeling of many. "Thinking that we were all raised with the Mercator map makes me feel kind of mise-ducated," she said, "because all these years we've been using the wrong map."

## Race as a Central Issue

As politically taboo topics became normalized in our class, race and racism became central discussion topics. Students talked and wrote about race and racism as easily as about field trips and homework. A number of my projects focused more closely than this one on race (see "Real-World Projects, page 195). However, because race was an ordinary topic of conversation and a political issue about which students were concerned, and because I asked which races (in general) lived in the North and the South, students also analyzed the project using race as a point of inquiry.

Several found the shrinking of the South relative to the North and the skewing of the relative sizes of Alaska and Mexico especially troublesome. Sandra had a particularly cogent analysis. "Doing this project has opened my eyes in different ways," she wrote. "I am learning how small details like maps, etc., have a lot to do with racism and power. Even though these kinds of things are small it can make a big difference in a person's view after learning what's really going on."

I also asked students to reflect on why the Mercator map is used so much in schools. One student, Elena, wrote in response, "I guess that's so because they wanted to teach [that] all Americans [are] superior and that all whites (that color and race) are better and superior than us (brown or lightly toasted, hardly white and Mexican). We were always taught that we were a minority and didn't deserve anything."

Issues of race and racism also surfaced in students' responses to the last question I asked: "In your opinion, is this in any way connected to anything else we've studied over the last two years?" Among those who answered affirmatively, Javier wrote, "It all has to do with white being the superior ... because the mapmaker was from Europe, he put Europe in the 'middle of the world' and on top of other countries. But

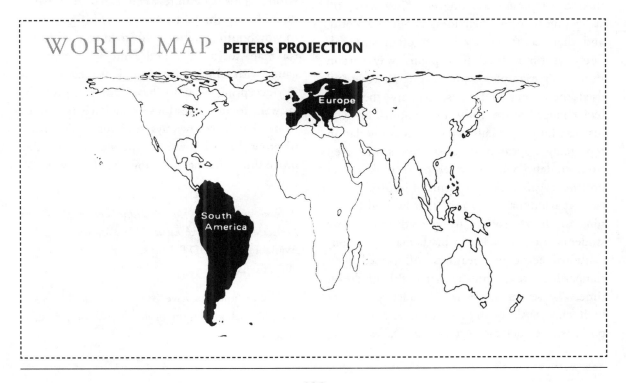

WORLD MAP **PETERS PROJECTION**

as we know, there is no such thing as a middle of the world."

Marisol also commented on race. "I think this is connected to what we studied over the past two years about equality, minorities, etc.," she wrote. "These maps seem to make minority-filled countries smaller than white-filled countries. To make us smaller and more insignificant? Yes, definitely."

## Reflections

A principal purpose of this and other projects was to get my students to think deeply about their educational experiences and what they had learned both in and out of school. I wanted my students to discover for themselves that "what you see is not always what you get." I did not want my students to take my word, nor anyone else's, without questioning. I wanted them both to use mathematics to read their worlds and to see the value of doing so—that is, to also develop a particular disposition towards using mathematics for social analysis.

My students did begin to read their world with mathematics, but clearly they were still developing in this area; for a variety of reasons, their development was uneven. They were still struggling to form their ideas about the world, and their contradictory beliefs often surfaced. Some at times tried to explain away certain things; for example, hypothesizing that we still used the Mercator because perhaps the earth had changed since 1569 and mapmaking was just catching up. This type of accommodation repeatedly appeared in class, sometimes even from students who at other times took strong positions against various forms of injustice.

It is important to note that beyond beginning to analyze society using mathematics, my students also succeeded in mathematics on conventional academic measures. All passed their standardized tests and graduated 8th grade on time, and several made it to academic magnet high schools. Although these are necessary accomplishments under the current high-stakes testing

policies, they are not enough. Educator Gloria Ladson-Billings' perspective on educating African American students also applies to working-class Latinos and other marginalized students. In her 1994 book *The Dreamkeepers,* she writes:

> [P]arents, teachers, and neighbors need to help arm African-American children with the knowledge, skills, and attitude needed to struggle successfully against oppression. These, more than test scores, more than high grade-point averages, are the critical features of education for African Americans. If students are to be equipped to struggle against racism, they need excellent skills from the basics of reading, writing, and math, to understanding history, thinking critically, solving problems, and making decisions.

A broader definition of educational success incorporates the vision of working for social justice. Mathematics can be an appropriate tool to realize this vision. The words of my student Lupe exhibit the sense of purpose needed in this struggle.

In commenting on our map project she wrote, "This is definitely connected to all we've done during the past two years. This goes back to why South America is so small in the Mercator map, to injustice and some sort of propaganda with false information. This relates to not just accepting what we have, but to search for answers to our questions. You have taught us to do that in many ways, and that only makes us grow. Who knows? Maybe we can someday prove things wrong and show the right way!" □

---

*A New View of the World: Handbook to the Peters Projection Map, 3rd Edition,* W. L. Kaiser 2005. Available from ODT Maps, http://odtmaps.com. 800-736-1293

Names of students have been changed. See www.rethinkingschools.org/math for actual project used in class.

# REAL-WORLD PROJECTS

## SEEING MATH ALL AROUND US

BY ERIC (RICO) GUTSTEIN

The following is a brief, chronological list of some of the projects done by my middle school students from 1997 to 2003. Each project has a short description, a summary of the mathematics involved, and an explanation of the writing the students had to do.

**1. WHO IS THE INS ARRESTING?**
Students read a newspaper article about the disproportionate number of Mexicans the INS was arresting, relative to the national-origin percentages of undocumented people. The mathematics included percentages, ratios, and proportions, and students also had to pose their own mathematical problems based on the article. Students had to write about their view of the INS policy.

**2. "MORNINGSIDE PARK TO BE PAVED FOR CONDO PARKING LOT."**
Students read a newspaper article about a developer who wanted city permission to raze a small park in order to build parking for his condominium development. They had to use two maps with different scales to determine distance and time between various locations. The mathematics also included ratios and measurement. Students had to write about whether the condo developer should be allowed to raze the park.

**3. GROWTH OF LATINO POPULATION IN THE METROPOLITAN AREA.**
Students read a newspaper article and had to use percentages and graphs to answer (and pose) several questions. They had to write about the social implications of the changing demographics.

**4. "TOMATO PICKERS TAKE ON THE GROWERS."**
Students read a *New York Times* article by this name, about how farmworkers' wages had gone down over 20 years, even without taking inflation into account. Mathematical ideas included COLA, absolute/relative comparisons, percentages, graphing, and inflation adjustments. Students had to write about what they would do if they were in the shoes of the workers.

**5. DEVELOPING SURVEYS (YEAR 1).**
Students developed and conducted surveys in small groups on mean-

ingful topics (i.e., serious and non-trivial). These included abortion (two groups), gay marriage, Darwin's theory of evolution, teen pregnancy, and discrimination against Latinos. Each group had to survey 80 to 100 people, make graphs, present findings, propose open research questions, write conclusions, and justify their choices along the way.

### 6.   WILL DEVELOPMENT BURY THE BARRIO?

Students read a long feature article about development in the school neighborhood. The mathematics included percentages, compound interest, ratios, and proportions. Students had to write about whether or not development would destroy the community.

### 7.   RACISM IN HOUSING DATA? (PART 1).

Students were given the 1997 median house price of the most expensive suburb in the area and asked if racism was a factor. Students had to describe and hypothesize data that would lead them to conclude that racism was a factor and data that would lead them to conclude racism was not a factor. Mathematics included data analysis, percentages, and problem-posing.

### 8.   SAT DATA.

Students were given 1997 SAT and ACT scores, disaggregated by race, gender, and income level. Students had to analyze correlations, create scatter plots, and write a letter to the Educational Testing Service raising questions and/or giving their opinions.

### 9.   WORLD WEALTH SIMULATION.

Students used data on wealth and population by continent to distribute cookies ("wealth") to people in the classroom ("continents"). Mathematics included determining percentages, absolute/relative comparisons, ratios, and proportions. Students had to write their thoughts on the distribution. (For a complete description of this activity, see page 89.)

### 10.   "WE ARE NOT A MINORITY."

Students looked at a photo of a billboard which showed Che Guevara with this statement. They had to write about how mathematics entered into the picture. After writing, the class analyzed world population by race and discovered that people of color constitute a large majority of the world's population.

**11.   WATCH AD.**

Students read a newspaper advertisement for a watch that cost $26,125. They had to write a creative piece about the watch and its price, using mathematics at the level of at least 8th grade, and have their opinion about the watch and ad come through in the piece.

**12.   FAIR ADVERTISING (BARNES & NOBLE AD).**

Students read a newspaper advertisement which used graphical representations to show the volume of books owned by Barnes & Noble and a competitor. Students had to define a "fair" ad, analyze the advertisement to determine whether it was fair, justify their decisions, and then, if they believed the ad to be unfair, create a new ad that was fair. Mathematics included measurement, geometry, percentages, ratios, proportions, and graphing.

**13.   U.S. WEALTH SIMULATION.**

Students used data on the wealthiest one percent, next 19 percent, and bottom 80 percent of the U.S. population to distribute cookies to themselves (representing the three groupings) by the wealth percentage for each group. Mathematics included percentages, absolute and relative comparisons, ratios, and data analysis. Students had to write their thoughts on the U.S. wealth distribution and how it compared to the world wealth distribution, and also whether using mathematics was necessary for them to understand the ideas. (For a description of a similar activity, see page 89.)

**14.   DEVELOPING SURVEYS (YEAR 2).**

Similar to year 1 but in greater depth.

**15.   PETERS VS. MERCATOR MAP PROJECTIONS.**

Students were given Mercator maps and the area for Mexico and had to find the areas of several other land masses. They then consulted almanacs to determine the real areas, which demonstrated that the visual representations of the Mercator projection were inaccurate. They then compared and contrasted the Peters and Mercator maps. Mathematics included measurement, geometry, ratios, and proportions. (For a complete description of this activity, see page 189.)

**16.   RACISM IN HOUSING DATA? (PART 2).**

Students analyzed data from almost 200 metropolitan suburbs to correlate house prices with racial composition of the suburbs. Mathematics included data analysis and graphing. They had to graph, use scatter plots, find correlations, and then write and justify whether or not they believed racism was a factor in their data and, if so, how and why.

### 17. THE MATHEMATICS OF RANDOM DRUG TESTING.

Students simulated a situation where school officials had a very accurate drug test and a small percentage of students were using drugs. They then looked at the probability that a randomly chosen student who tested positive was really using drugs. Adapted from the IMP curriculum. Mathematics included probability, expected value, randomness, and statistical reliability.

### 18. DRIVING WHILE BROWN/DRIVING WHILE BLACK.

Students simulated racial profiling data of "random" traffic stops, then compared the results of their simulations to actual data. Students analyzed the data to decide whether racial profiling was a real problem and, if they believed it was, proposed what should be done about it. Mathematics included probability, data analysis, and rational numbers. (For a complete description of this activity, see page 16.)

### 19. B2 BOMBER PROJECT.

Students compared the cost of one B2 bomber to the cost of four-year scholarships to a prestigious out-of-state university for all the graduating seniors in the neighborhood high school, as well as all of Chicago. Mathematics included using large numbers, scientific notation, and ratios. (See the related activity, page 110.)

### 20. MATHEMATICS OF DEMOCRACY.

Students read a cartoon (at right) about public opinion on the Iraq War. They then wrote articles or editorials about the cartoon and analyzed how the cartoonist used mathematics to make his point. Mathematics included percentages, graphs, ratios, and proportions.

### 21. MORTGAGE LOANS—IS RACISM A FACTOR?

Students read a newspaper article documenting the disparity between African Americans, Latinos, and whites in securing home loans in the Chicago area. Students analyzed the data, solved a number of problems, and wrote essays arguing whether racism was a factor in the disparities and using mathematics to justify their views. Mathematics included algebra, percentages, graphs, data analysis, and ratios. (For a complete description of this activity, see page 61.) ∎

## DO YOU SUPPORT INVADING IRAQ?

### CARTOON TEACHING SUGGESTIONS

Show students the cartoon and ask, "In the cartoon, why did the responses to the survey change? What does this tell us about the importance of how questions are constructed?

What does the cartoon imply about President Bush's concern for public opinion? See Mathematics of Democracy in "Real-World Projects," at left, for other ways to use this cartoon.

# DECONSTRUCTING BARBIE

## MATH AND POPULAR CULTURE

BY SWAPNA MUKHOPADHYAY

Math and popular culture are rarely associated, although phrases like "do the math" are commonly heard. Undoubtedly this is a prime example of the unfortunate disconnect between math in school and its real-life applications. Many teachers, however, use mathematics for the analysis of complex real-world issues. Teaching mathematics as a cognitive tool for everyday sense-making is an approach that addresses the alienation of many students from early grades.

Beyond sense-making, using mathematics as a tool to interrogate issues of importance to students, their communities, and society in general brings to light the generally hidden cultural, historical, and political nature of mathematics education.

Here's an activity that treats mathematics as a tool for learning about a number of social issues. It begins with inviting a Barbie doll—the popular cultural icon—to the classroom. The activity teaches concepts such as averages and ratios, in a way suitable for middle school students.

### MATERIALS

Besides a few Barbie dolls from my thrift-store collection, I also bring tape measures, a ball of string, and calculators as my material props.

### PROCEDURE

The task starts with a probe: What would Barbie look like if she were as big as you? Students, working in small groups, need to figure out first the "average" of the group, so that they can begin to "construct" a real-life Barbie of the same height. After confronting the mathematical notion of average and its contradiction to real life (for example, "Jenny," the "average" in our group, is a blond European American, whereas I am a dark-skinned Indian), the students measure the ratio of the height of the designated person to that of the Barbie doll to find the enlargement ratio or the scaling factor.

Then, with a table I provide, they measure the body parts of the doll and compute its corresponding real-life measurements.

The next part of the task consists of drawing the contour of Jenny-the-average on butcher paper with one color marker, and then superimposing the blown-up Barbie on Jenny's image with a different color. Drawing two human figures—the contour of familiar Jenny and Barbie-as-big-as-Jenny—immediately highlights the unrealistic

# DECONSTRUCTING BARBIE

## MEASURE UP, BARBIE!

$$\text{scaling factor} = \frac{\text{height of person}}{\text{height of Barbie}}$$

### BODY MEASUREMENTS

|  | BARBIE | BARBIE SCALED | REAL PERSON | RATIO |
|---|---|---|---|---|
| BUST |  |  |  |  |
| WAIST |  |  |  |  |
| HIPS |  |  |  |  |
|  |  |  |  |  |
|  |  |  |  |  |
|  |  |  |  |  |

### OUTLINE MEASUREMENTS

|  | BARBIE | BARBIE SCALED | REAL PERSON | RATIO |
|---|---|---|---|---|
| LEGS |  |  |  |  |
| ARMS |  |  |  |  |
| HEAD |  |  |  |  |
| NECK |  |  |  |  |
|  |  |  |  |  |
|  |  |  |  |  |
|  |  |  |  |  |
|  |  |  |  |  |
|  |  |  |  |  |

dimensions of Barbie, which are not so obvious on the doll itself, or even in the computed measures collected in tabular form.

Repeating the exercise with male superhero action figures and the "average" male student makes it clear that the issue is not confined to female representations.

### DISCUSSION

This activity then generates discussion connecting to the impact of representations in popular culture on body image, self-worth, and eating disorders. Other issues raised include the superficially "multicultural" nature of contemporary Barbie and the sweatshop labor that produces the dolls. ■

# CHAPTER TWENTY-SIX

# Multicultural Math

## One Road to the Goal of Mathematics for All

BY CLAUDIA ZASLAVSKY

As I look back on my teaching experiences in Greenburgh Central Seven, a school district just north of New York City that was known nationwide for its decision in 1951 to integrate its schools through busing, I am struck by the similarity between the problems we encountered in the 1960s and 1970s and those of today.

The solutions we worked out may be of value to teachers trying to reach all students with mathematics. In particular, this article will deal with the introduction of elements of a multicultural, anti-racist mathematics curriculum in the middle and secondary grades.

The National Council of Teachers of Mathematics recognizes the relevance of culture, saying: "Mathematics is one of the greatest cultural and intellectual achievements of humankind, and citizens should develop an appreciation and understanding of that achievement, including its aesthetic and even recreational aspects." I agree. Multicultural education should include the contributions of all peoples, as well as concern for their problems and difficulties. Further, it should embody the consideration of the factors in our society that prevent the effective participation of all members of society.

The '60s was a period of ferment and change. One effect on our school district was the students' demand for courses in African history and Swahili, courses that were relevant to their culture. The faculty, too, was offered an optional course in African history, a subject that generally had received inadequate, if any, treatment in the college curriculum.

For my term paper in the course, I decided to research the development of mathematics in Africa south of the Sahara, but I was amazed to discover that there was no heading in the library catalogs for African mathematics or for any of its branches.

Mathematical practices and concepts arose out of the needs and interests of people in all societies, in all parts of the world, in all eras. All peoples have invented mathematical ideas to deal with such activities as counting, measuring, locating, designing, and, yes, playing, with corresponding vocabulary and symbols to communicate their ideas to others. I was convinced that African peoples were no exception and that such information must be available. My research eventually led to the publication of a book on the subject: *Africa Counts: Number and Pattern in African Cultures* (see Resources, page 264).

## A New Secondary School Mathematics Curriculum

Although our secondary school had few dropouts—partly because of the availability of an adequate number of guidance counselors, who stayed with the same students throughout their high school careers—we found that many students were doing poorly in the traditional academic mathematics courses and were dropping the subject as soon as they had fulfilled the minimum requirements. These students were predominantly of working-class backgrounds, both black and white. This was not an acceptable development in a district that prided itself on offering the best possible education to all its students.

Like today, the '60s was a period of great concern about the status of mathematics education in this country. The Soviet "conquest" of space with the launching of Sputnik in 1957 sent shock waves through our government and business establishment, resulting in increased funding to upgrade mathematics and science education. Our district, ever on the alert for such opportunities, obtained funds to write a new curriculum to attract potential math dropouts. We were fortunate, too, in that the community supported us as we increased the number of courses offered by the mathematics department, a number exceeding what was made available in some larger and wealthier districts nearby.

Our mathematics department developed a complete curriculum for grades 9 to 12, designed to be user-friendly and combining hands-on activities with many of the topics usually covered in grades 9 to 11. The 12th-grade unit was based on elementary probability and statistics. The entire curriculum included more computation than is usually found in academic mathematics courses at this level, presented in a manner that would enable students to understand the underlying concepts and some aspects of number theory. As much as possible, applications were derived from local issues and the real world.

I will describe briefly the most relevant topics in the 12th-year course, then discuss more fully one of the most successful projects, involving census data for the local area. The topics listed below were not necessarily included every time the course was given: Flexibility was a great advantage in designing our own courses. We could add what seemed relevant at the time and omit the topics that did not work well. The topics included the following:

- Probability theory, involving experiments with coins, dice, thumbtacks, and playing cards and a discussion of mortality tables, lotteries, and heredity.
- Statistical distributions based on students' data: heights, birthdays, number of children in the family.
- Descriptive statistics (tables and graphs) relating to such topics as the integration of whites and "non-whites" in housing; the amount of nicotine in different brands of cigarettes; the lifetime dollar worth of an education; the comparison of education and earnings of males and females; and the comparison of blacks and whites with regard to education, income, infant mortality, life expectancy, and other factors.
- Computation, significance, and applications of the mean, median, and mode; histograms and the normal distribution curve.
- Computation of grade-point averages, the consumer price index, and the cost-of-living increase incorporated into the contract of the United Automobile Workers union.

Certain topics evoked a great deal of interest. This was the period of the Vietnam War, and young men were very concerned about their futures. A Department of Defense publication showed that in 1967 black men composed nine percent of the armed forces, but 11 percent of all men assigned to Vietnam and 15 percent of the deaths in Vietnam. The students discussed the reasons for the discrepancy, illustrating their arguments with stories about friends and relatives.

Another interesting set of statistics involved the difference between the cost of a shopping basket of food in the low-income and middle-class neighborhoods of New York City: Poor people paid more for their food than higher-income people! Langston Hughes' lines about Harlem from his poem "The Panther and the Leash" were appropriate: "Uptown on Lenox Avenue / Where a nickel costs a dime … ."

## Analysis of Data About the Community

The most empowering and, therefore, the most successful lessons were related to the students' own community. Using the Department of Commerce publication *Standard Metropolitan Statistical Areas Outside of New York City, 1960 Census,* I drew a map showing the five standard metropolitan statistical areas (SMSAs) that composed the school district. I leafed through the book, selecting the features that I thought would be of greatest interest to the students, and copied the data for each SMSA. Among the categories I used were these:

- Total population, classified as white, black, other.
- Education and median family income for each category.
- Male and female employment and unemployment rates for whites and non-whites.
- Housing—owned or rented, condition, value; number of persons per room—for whites and non-whites.
- Automobiles owned.

Every student received a copy of the map and the data. After some discussion, the students paired off, each couple taking responsibility for comparing the five SMSAs from the point of view of two or three characteristics. We had some general discussion about making tables and graphs to display the data, and then they got to work. Although they worked in pairs, each student was responsible for producing the tables and graphs.

(I had assumed that this was a simple assignment, but I was wrong. For some, planning the scale for the graph was a formidable task. Their first attempts revealed how little they understood. Their scales did not fit on the paper. They did not know how to label the intervals after they had marked them on the vertical axis; if the data gave median incomes of $2,320, $4,350, $5,489, $4,190, and $3,085 for the five SMSAs, those were the numbers placed next to the marks on the vertical line. Several students were successful only after three or four attempts.)

Then the class came together to compare and analyze their productions. With the boundaries of each SMSA clearly marked on the map, each student knew exactly where he or she fit into the total picture. It was evident that the predominantly white, middle-class population had far higher incomes, better housing, more education, and all the other attributes of a higher standard of living than the more poorly endowed black and working-class population. Of course, the students had been aware of these differences before. But here were the numbers, and numbers don't lie. Here were the graphs, and a picture is worth a thousand words.

Most of the students were taking sociology as their senior-year social studies course. The social studies department considered me an honorary member of their department, and they welcomed the input from my classes as a basis for further discussion in the sociology course. Nor did parents object to our tackling such a controversial subject as societal inequities. In a district that had opted for school desegregation, we could generally count on the support of the majority of the community in dealing with such topics. Many families, both black and white, had chosen to live in this community precisely because it was less segregated than most of the surrounding area.

Both the faculty and the students considered these courses successful in overcoming math anxiety and math avoidance. We tried to eliminate the need for remedial courses at the high school and college levels by incorporating a review and by reteaching, if necessary, concepts from arithmetic in a context that enhanced the students' self-esteem, rather than labeling them "dumb" and increasing their fear of mathematics. Furthermore, these topics in arithmetic took on more meaning than if they had been studied and drilled out of context.

Many of these potential dropouts from mathematics continued to take mathematics courses throughout their high school years. We were particularly successful with blacks and working-class young people of all races, those groups that had been underserved and underrepresented in mathematics. Young women were not a problem; females and males in our district were equally represented in advanced mathematics classes as far as the pre-calculus level. But we did make conscious efforts to encourage all students, especially "minorities," to pursue mathematics.

## Mathematics comes alive when students participate in activities that illustrate how mathematical decisions arose from the basic needs of societies.

### The Mathematics of Various Cultures

In the early '70s I began to explore the possibilities of introducing into the curriculum the study of mathematics as a cultural product, the kinds of mathematics developed by various societies to satisfy their own needs. My focus at the time was on societies in Africa south of the Sahara; later I broadened my scope to include some of the underrepresented societies throughout the world. In recent years the mathematical and scientific developments of non-European cultures have become the focus of study for a number of scholars. Not to include such contributions is to imply that these people had no mathematics or science. The same applies to women of all races and ethnic backgrounds.

I took advantage of a sabbatical year to develop and test materials for several grade levels, based on the infusion of African mathematical ideas into the curriculum.

In our district the 9th-grade social studies curriculum included the study of Africa. The two teachers who team-taught this large, heterogeneous social studies class were perfectly happy to allow me to take over for a week. On the first day I showed slides illustrating mathematical practices that several African peoples had developed in the course of their everyday activities. The following day I introduced a unit about networks, on which we spent the remainder of the week.

From *Africa Counts: Number and Pattern in African Cultures*

A DIAGRAM OF A CHOKWE NETWORK SIMILAR TO THOSE DRAWN IN SAND.

A network consists of a set of points (vertices) and the paths (edges) connecting them. My favorite network is a figure that was drawn in the sand by elders of the Chokwe people of Angola and Zaire to illustrate their myth about the beginning of the world (see diagram above).

These African networks are examples of mathematical graphs. Graph theory is a growing field of mathematics and has important applications in modern technology relating to communications, highways, and so on. The task I set the students was to trace the given networks and to determine which designs were traceable: that is, could be drawn in one sweep of the pencil, without either lifting the pencil or going over the same line segment more than once. Then I challenged the students to determine the necessary conditions for traceability.

The project was open-ended; some students were able to trace only the given networks; others drew traceable networks of their own invention, and still others tried to respond to the theoretical challenge. Most students indicated on their evaluation sheets that this was a positive experience for all, including the teachers, who later told me that several of these students had not otherwise participated in class since the beginning of the school year. They had rejected the papers I distributed the first day, but when they saw their classmates engaged in hands-on work and animated discussions, they, too, wanted to participate.

It seemed to me that, in general, this infusion of the cultural aspects into the curriculum was successful. African American students, in particular, found such materials relevant to their lives and backgrounds. One young woman chose to instruct an 8th-grade class in African applications of network theory as her senior year project.

Many mathematical topics lend themselves to the infusion of cultural applications. In the process of designing a rug similar to those woven by Navajo women, students deal with symmetry, geometry, and measurement. At the same time they learn to appreciate the degree of mathematical knowledge required to weave complex designs, generally with no pattern and no specific instructions to follow. With this understanding, "women's work" takes on new significance. To mention just one problem: How does the weaver compute the number of rows required to make the border pattern "come out right" when she reaches the far side of the rug?

Mathematics comes alive when students participate in activities that illustrate how mathematical decisions arose from the basic needs of societies. For example, why do people build their homes in certain shapes and sizes and use particular materials? An investigation into the building styles in various cultures is a valuable source of experiences with shapes and sizes, perimeter and area, estimation and approximation, while at the same time it shows the relevance of mathematics to social studies, art, and other subjects. Students might consider a tipi, an African mud-and-wattle roundhouse, or an Inuit igloo a "primitive" dwelling compared with an urban apartment house or suburban ranch house. Yet the people who build these homes, or built them in years past, are using their available materials and technology to the best advantage.

## Relevant, Not Trivial, Multiculturalism

A word of caution: Teachers must be careful that they do not introduce cultural applications as examples of "quaint customs" or "primitive practices." These applications must form an integral part of the mathematics curriculum. They must inspire students to think critically about the reasons for these practices and to dig deeply into the lives and environment of the people involved. It is so easy to trivialize the concept of multicultural education by throwing in a few examples as holidays approach. Better not to do it at all!

The multicultural context is relevant to many aspects of the mathematics curriculum. A discussion of the number words and numeration systems of non-English-speaking peoples may

do wonders in raising the self-esteem of students who speak these languages, while also enhancing the understanding of all students. It may come as a surprise that in some languages, numeric grouping is done by 20s rather than by 10s as in English. Games of chance, games of skill, and patterns in art and architecture are all sources of learning experiences. Some of the richest contributions may come from students in the class and their families.

Bringing the world into the mathematics class by introducing both cultural applications and current societal issues motivates and empowers students. Such mathematical content offers opportunities for project work, cooperative learning, connections with other subject areas, and community involvement. To carry out such a program effectively requires a new approach to curriculum development, teacher education, and assessment processes. But it is well worth the effort. □

This article is adapted from a version that originally appeared in *Reaching All Students with Mathematics,* Gilbert Cuevas and Mark Driscoll, eds. (NCTM, 1993).

# TRACKING PA ANNOUNCEMENTS

## COLLECTING AND ANALYZING DATA IN THE CLASSROOM

BY ROBERT M. BERKMAN

Several years ago I was hired to teach in a public middle school located in a blue-collar section of Brooklyn. There was an almost constant flow of announcements over the public address system, to the point where a 40-minute class might be interrupted half a dozen times.

Fueled by defiance and exasperation, I decided to reserve a corner of my chalkboard to tick off each interruption. My students were confused when I abruptly and silently stopped a lesson to add the latest announcement to my tally, but after the third or fourth time, they quickly put together the cause and effect. Their reaction was to chuckle. Apparently they were so conditioned to this constant stream of blather that they were surprised that it bothered me at all. I remember one student laughing as I counted off what must have been the 10th announcement of that day, remarking: "Yo, Mr. Berkman, that sh–– be driving you crazy!"

### COLLECTING THE DATA

After collecting a week's worth of tally marks, my students were not surprised that the total was quite high (as many as 25 announcements were made in a single day). Yet they felt that what I was doing had grown, to use their word,

"lame." I told them that I had a theory that there were certain times when the PA system was most used, and that perhaps I could prepare myself for those times (by not attempting to actually teach a lesson).

A spirited discussion ensued: One student said it was always at the end of the day, when kids had to leave early, while others thought it was in the morning, when the principal delivered his daily pep talk. I refined my corner of the board to include eight boxes, each representing a different period of the day.

That week my students were careful to make sure I kept track of all announcements, popping in between periods to ask: "Hey, Mr. Berkman, I heard two last period. Did you get them?" We looked at the daily totals, eventually creating a histogram for the entire week. Our results showed that most announcements occurred during the middle of the day, in fact, most often during lunch periods, when teachers were either being reminded to show up at lunch duty or kids were being sent to the office for some trouble they had caused.

During the next few weeks, we kept track of announcements in a variety of ways. We set up a chart classifying announcements by topic, including "general interest," "requests for a teacher," "requests for a student," and "miscellaneous." Our research found that while "general interest" announcements occurred at the beginning and end of the day, the "student requests" happened during lunch periods, while "teacher requests" and "miscellaneous" occurred at random times.

### HISTOGRAMS AND PIE CHARTS

By this time, the students were collecting their own data, to the point that they would correct the information I had carefully recorded. After a discussion about the announcements, the students grumbled about the different people who made them, noting that it was usually the principal, his secretary, or one of the three assistant principals who was the "guilty" party. To test their theory, my students kept charts on who was making announcements, what time of day they were made, and the subject of each. They tabulated this data and made a variety of histograms, pie charts, and line graphs.

When the students asked what we would do with all the information, I jokingly suggested that we write up our conclusions and submit the results of our investigation to the principal in an effort to make him aware of how many interruptions were taking place during the day. As a new teacher, I didn't seriously consider taking this action. It was hardly necessary: Within a few days, other teachers had noticed their students tracking announcements in their notebooks, which soon came to the attention of the assistant principals.

### TAKING ON SCHOOL CULTURE

During the next week the steady flow of interruptions slowed to a trickle—although they did stage a comeback after a month or two. (Was this a sociological example of the Heisenberg Uncertainty Principle, that the act of recording these interruptions actually influenced how many were made?)

I cite this experience because what began as an angry rejection of school culture was transformed into a subject for data collection, evaluation, and analysis. The students took interest in this because it was something that annoyed them day in and day out, but they didn't regard it as something that was worthy of collection and study, for they were so enmeshed in the school culture that it didn't occur to them to question it.

Collecting data got them involved in thinking about why things were the way they were, and as they answered one question, others sprang up, which ultimately led them to the realization that things didn't have to be this way: Our school could function just as effectively without all these interruptions; furthermore, it was the adults who had to mend their ways, not the kids. ■

## PART 3

David McLimans

# Infusing
# Social Justice Math
# into Other Curricular
# Areas

# CHAPTER TWENTY-SEVEN

# Ten Chairs of Inequality

BY POLLY KELLOGG

Inequalities of wealth are becoming more extreme in the United States. While billionaires double their wealth every three to five years, we have by far the highest poverty rate in the industrialized world. No industrialized country has a more skewed distribution of wealth. Students need information about this concentration of wealth—and the power that accompanies it—in order to become critical thinkers and aware citizens.

A Boston-based group, United for a Fair Economy, has developed a simulation activity to dramatize the increasingly unequal distribution of wealth. I describe here how my college human relations classes respond to the exercise. It can easily be adapted for younger students.

To begin the simulation, I ask 10 students to volunteer to line up at the front of the room, seated in their chairs and facing the rest of the class. I explain that each chair represents 10 percent of the wealth in the United States and each occupant represents 10 percent of the population; thus when each chair is occupied by one student, the wealth is evenly distributed. I explain that wealth is what you own: your stereo, the part of your house and car that are paid off, savings like stocks and bonds, vacation homes, any companies you own, your yachts, villas on the Riviera, private jet airplanes, etc. Then I ask students to estimate how much wealth each family would have if the wealth were equally distributed. Students usually guess about $50,000 and are surprised to hear that the answer is $250,000.

I ask them what it would feel like if every family could have a $100,000 home, a $10,000 car paid for and $140,000 in savings. Some make comments like, "It'd be wonderful. I wouldn't have to work two jobs and take out a loan to go to college." But many can't imagine such a society. Others express concern that the incentive to work would be taken away: "It sounds like socialism."

I invite the class to speculate how many chairs belong to the richest student, whom I will call Sue. In 1998, the richest 10 percent owned 71 percent of the wealth—thus Sue controls about seven chairs. I tell the six students sitting nearest to Sue to give up their chairs to her and move to the poorer end of the lineup. I provoke them by telling them that the standing students can sit on the laps of the three students seated at the end, and I invite Sue to sit in the middle of her seven chairs, to stretch out, relax, or even lie down across the chairs.

I then announce that Sue's arm represents the wealthiest one percent of families and that her arm's share of the wealth doubled from two chairs (22 percent of the wealth) to four chairs

I tell those worried about socialism that they have nothing to fear. We have nowhere near an equal distribution of wealth in this country. The poorest 20 percent of the population is in debt, and the next 30 percent averages only $5,000 in wealth (primarily in home equity).

I ask students at either end of the lineup which one of them wants to represent the richest 10 percent and experience being rich. Some students volunteer happily and others express distaste at the idea. When asked about their motives, they say, "I'll never be rich, so I'd like to see what it feels like," or "I don't want to oppress other people, and rich people exploit their workers." Sometimes a student, often female, will say, "I don't like to be above other people."

(38 percent) from 1979 to 1998. I solicitously help Sue find a comfortable position with one arm stretched over four chairs. To engage Sue in clowning and playing up her role, I offer her food or drink.

## "I Want a Revolution"

I ask the other nine students crowded around three chairs what life is like at their end of the line. "We're pissed and tired of working all the time," is a typical comment. Another is, "I want a revolution." I ask students if, in real life, they or people they know are crowded into the bottom one or two chairs, and what that's like. Working-class students tell stories of financial stress they have experienced, such as, "My mother had to work at two jobs to support us."

"My family was really poor when my dad was laid off. We lived on macaroni and cheese." Often one student, usually a white male, says he has hopes that he can work hard and join Sue.

Students' knowledge of how inequality is rationalized erupts when I ask, "What do those in power tell us to justify this dramatic inequality?" Typical student answers are: "They work harder than we do." "They create jobs." "The United States stands for equality and justice for all." "It's our fault if we don't make it." If students do not mention scapegoating, I bring it up. I may select one student to represent the poorest 10 percent and ask, "Wouldn't there be more money for the rest of you if he or she weren't ripping off the system for welfare?" I also ask, "Who does Sue want you to blame for your tough economic conditions?" Answers range from welfare moms and immigrants, to gays and lesbians, to bad schools.

When I ask the nine students grouped around the three chairs why they don't get organized to force a redistribution of the wealth, they offer a variety of answers. "We're too busy working to organize." "We are told we can't change things." "We don't get along with each other." "They'd call out the army to stop us."

At some point I ask the class to describe the "super rich"—the one percent, Sue's arm. College students share examples from their experiences. A junior high coach described a local billionaire who offered to write the coach's

school a check of any amount in order to get his child on the baseball team. Another student worked as a waiter in an elite club where "you had to be elected to be able to have lunch there." The club was all white and only recently began allowing women on the premises; some of the older men refused to let her serve them because they resented her presence. Another student described doing carpentry in a mansion of the DuPonts, "The faucet in the kid's bathroom cost $3,000."

The athletes and entertainers whose salaries are hyped in the media and newly rich entrepreneurs like Donald Trump and Bill Gates are always mentioned. I point out that these are the upwardly mobile people, who moved from the three chairs up to Sue's chair. How often does this happen? Why do we hear so much about them? I want students to understand that the exaggerated publicity about these rags-to-riches icons perpetuates the myth that anyone who tries can make it.

Most texts and teachers stop after they have taught students about the unequal distribution of wealth, but that is only a piece of the picture. We need to go on to ask why wealth is so unequally distributed. Where does wealth come from? Why does our system concentrate wealth in the hands of so few? And what can ordinary people do to effect change? The simulation creates a foundation for these later lessons. □

Illustration: United for a Fair Economy

# FAST FOODS, JUNK FOODS, AND ADVERTISING

## ANALYZING THE BARRAGE OF ADVERTISING AIMED AT CHILDREN

BY BOB PETERSON

In 2001 the U.S. Surgeon General announced that obesity had reached epidemic levels in this country. In 2012, the Centers for Disease Control and Prevention said, "Childhood obesity has more than doubled in children and tripled in adolescents in the past 30 years." While there are several causes of this epidemic, one clear aspect is the constant barrage of advertising students receive encouraging them to eat junk food.

Juliet Schor, in her book *Born to Buy,* provides a wealth of data and analysis of what she calls the "corporate-constructed childhood."

Her book is filled with data that could be used by math teachers to develop lessons that students would directly relate to and benefit from. Two of those lessons are included here.

### LESSON 1

A study, published in the *Journal of the American Dietetic Association* (April 2008), examined 27.5 hours of programming for children on Saturday mornings in May 2005. They found "49% of advertisements shown were for food (281 food advertisements out of 572 total advertisements). The most commonly advertised food categories were ready-to-eat breakfast cereal and cereal bars (27%), restaurants (19%), and snack foods (18%). Ninety-one percent of food advertisements were for foods or beverages high in fat, sodium, or added sugars or were low in nutrients. Cartoon characters were used in 74% of food advertisements, and toys or other giveaways were used in 26% of food advertisements."

### ACTIVITY

As a homework assignment, teachers can provide students with a chart for tabulating, in a set period of time, the number of commercials, product types, and brand names they are exposed to. Together the class can combine the data or do a separate analysis. Depending on the level, students could make charts and graphs, convert data to percentages and fractions, and contrast their findings with national data.

## LESSON 2

According to Schor, "An accounting from the May 2002 issue of Nickelodeon magazine, the most popular kid magazine in the country, found that of its 24 ad pages, 18 and a half were for junk foods. Five and a half were for candy and gum, and the remaining 13 were devoted to products such as Pop Tarts, Tang, Jell-O, and sugary cereals."

### ACTIVITY

Teachers can locate popular children's magazines and have students tabulate, graph, and analyze the nature of advertisements in those magazines. Again depending on the math level, students can display their data in different ways.

### QUESTIONS FOR CLASS DISCUSSION

1.   What are children learning from these advertisements?

2.   Why aren't there advertisements/commercials for fruits and vegetables?

3.   What influence do you think these advertisements have on your younger brothers and sisters? What would happen if your brothers and sisters only followed the advice of these advertisements?

4.   In Sweden, these kinds of advertisements are against the law, as people there think that children are too easily influenced to buy things that are unhealthy for them. Do you think that such advertisements should continue here in this country? Why or why not?

5.   How does using math help us better understand what is happening in our community, on TV, in magazines, and to our health?

### OTHER ACTIVITIES

•   Students could research the nutritional content of some of the advertised products.

•   Students could develop presentations on the impact of such advertising and present them to younger students, other teachers, or parent organizations.

•   Students could imagine what would happen if a person just followed the advice of the advertisers, and draw and write about this idea. Where appropriate, teachers could show selections of the documentary movie *Supersize Me* by Morgan Spurlock, which follows Spurlock's misadventures as he endeavors to eat only McDonald's food. ∎

# CHAPTER TWENTY-EIGHT

# Transparency of Water

## A Workshop on Math, Water, and Justice in Chicago

BY SELENE GONZALEZ-CARILLO
AND MARTHA MERSON

I f you are an environmental orga-
nizer, like Selene, your classroom is
a conference room, the community
garden, a church parking lot, a vacant apartment. The teaching starts after coffee
and greeting friends. It ends when participants feel work or family obligations
are too pressing to ignore.

Your students are everyone—from toddlers to the elderly; they come with a
variety of levels of formal education.

Your goals are to increase environmental justice, community well-being, and
individuals' health. You want fresh air and safe drinking water for all, for the
sake of the planet.

Such changes won't result from one workshop. Yet one workshop can open
minds, shift assumptions, and form connections that lay the foundation for collective action.

If you are a math educator/curriculum writer with an interest in data, as
Martha is, you teach in adult ed and K-12 classrooms, libraries, and living
rooms—anywhere you can sneak in math.

The teaching starts with a provocative statistic or a document with unfathomable numbers. Your students are often math averse. They are adults who will pause to think about size or scale, what the data convey.

Their motivation could be to earn a credential like a high school diploma, or simply to learn more. As one community member observed: "It's good to learn. I'm not dead yet." Many have plans to put learning to use in their communities, churches, and families.

Your goals are to encourage adults and youth to take a new look at numbers, to ask questions.

Statistics for Action (SfA), a National Science Foundation-funded project, brought us together. SfA prompted Selene to include activities to build numeracy with youth and adults. The priorities for the environmental justice movement fit well with SfA's goal of providing organizers and community members with tools and resources to use data for decision-making.

SfA offers many resources to build participants' familiarity with the literacy and numeracy demands of environmental quality reports so people who want to advocate and discuss local issues don't have to teach themselves everything from scratch. During the project, we led an SfA-inspired two-hour workshop in Spanish designed to probe adults' distrust of tap water and arm them with tools and knowledge to take on water quality/delivery issues.

## Do You Drink Chicago's Tap Water?

We made coffee and set up a water taste test. At 9 a.m., volunteers recruited for a six-week Leadership Program filed in for their fourth session. Among the participants were Carmela, a single mother, college student, and intern at Little Village Environmental Justice Organization (LVEJO); Elena, studying for a commercial trucker's license; and Sandra and Juan, who contribute labor and know-how to LVEJO's garden projects. The eight participants ranged in age from 10 to 60. Their project facilitator,

Norma, also a neighborhood resident, has a long history of social justice work.

### Drinking Water Taste Test/ Use Your Senses

We introduced the agenda by recapping the previous session.

Last time you talked about communication and opinions. This continues that discussion; it's a chance to discuss how combining stories and numbers make a powerful message. Today we will:

- Begin with a taste test of Mountain Spring bottled water, filtered water, and tap water.
- Watch excerpts of the video *The Story of Bottled Water.*
- Create messages with statistics.
- Examine Chicago's municipal water report.

Selene instructed the participants to sample water from each of three pitchers. They examined the water in their cups, swished it around, and swallowed.

Carmela: This is good, because I'm dehydrated.

Sandra: I'm a dummy; they taste the same to me.

Carmela: It's all the same water, I get it [then laughs], is it?

Elena: It's between A and B.

The participants reported their choices. Though divided in their preferences, most surprising to them was their difficulty in discriminating between bottled and tap water. Martha then asked:

How often do you buy bottled water?

We wanted to understand why and to what extent the participants paid for bottled water. Also, we hoped that the reasons they shared would be examined in *The Story of Bottled Water* (available at storyofstuff.org), which we had queued up.

Yoana: I buy bottled water because it's cleaner.

Carmela related what happened when her girlfriend offered her water.

> ... I'm like, "What are you doing!" because that's tap water—forget about it. [But] the tap water tasted better and I would tell my mama, "Drink water from the tap, it tastes very good," and she would say to me "No! Are you crazy? It has chemicals. ... Go back to Chicago."

We didn't judge participants' choices. Distrust of tap water runs deep in communities of color. In a 2011 study published in *Pediatrics & Adolescent Medicine*, of the parents surveyed, "Minority parents were more likely to exclusively give bottled water to their children."[1] Reasons cited included taste and safety. Researchers reported that parents of color were more likely to believe that bottled water has minerals and nutrients that tap water does not.

Little Village, where the participants live, is home to more than 90,000 residents, nearly half of whom are immigrants, according to the University of Chicago. Many come from places with a history of serious water issues.

In 2008 the Environmental Working Group published results of their investigation of 10 popular brands in nine states and found a total of "38 chemical pollutants ... with an average of eight contaminants in each brand," including industrial solvents and fertilizer residue. Consequently, the reliance on bottled water in communities of color is a serious concern, because it means that these communities are disproportionately at risk for health consequences associated with the use of bottled water.[2]

### Selling the Public on Bottled Water

Selene introduced the group to the video, which illustrates—with line drawing animations—the trend away from tap water and toward bottled water, situating this shift in a larger context of manufactured demand.

If you lived here 20 years ago, you would not see bottled water as much. How did so many people's opinions change?

Annie Leonard, the film's narrator, throws out intriguing statistics. In her commonsense way, she explains how bottled water is a problematic product along its whole life cycle: It takes petroleum to make the bottles, and we know how devastating oil spills can be; it takes fuel to move the bottles around and that pollutes the air; and then the bottles end up in trash cans or by the side of the road. Of those that make it to recycling centers, only a few are actually recycled. Leonard's message is simple: "We're trashing the planet; we're trashing each other; and we're not even having fun." Though ads may promise glamour and bottled water manufacturers say their product meets consumer demand, in taste tests across the nation, people chose tap water over bottled water. The animation of this point, a stick figure spitting out bottled water, adds comic relief to the disturbing story Leonard spins.

Two minutes into the video, at 2:05, we paused to let this fact sink in: Bottled water costs 2,000 times more than the cost of tap water. At 3:46 we heard and captured this: One third of bottled water sold in the United States is filtered tap water. At 4:42 we paused to record: 80 percent of plastic water bottles go to landfills.

Leonard goes on to explain how all of this got started by defining "Manufactured Demand" as the force that drives the production of goods. "In order to grow you have to sell stuff." When soft drink companies feared a drop-off in business, they fabricated the need for bottled water by inducing fear of tap water, making bottled water look seductive, and misleading the consumer. Marketed as a beverage, bottled water companies avoid the rigorous testing the Environmental Protection Agency requires of municipal water systems.

In the final minutes of the film Annie sends out the battle cry: Take Back the Tap!

She makes clear that viewers can take action and make a difference to save our right to clean water. After viewing the video, we asked: Was there anything surprising? Sandra spoke up:

> Sandra: I did not think that the bottles would be [piled up] ... just like that, mountains and mountains [of bottles] ... I used to consider myself "damned" because I could not buy this type of water. A friend would tell me ... "You're drinking dirty water from the tap!" But tap water isn't bad; they're just charging us double.

That was an important point. We wanted participants to create sound bites drawing on one of the three statistics and their experiences, imagining that they were talking to a neighbor who hadn't seen the film. Selene planned to elicit an example, stating a fact with a fraction, percent, and a ratio.

Selene: Instead of a third, what's the percentage?

Elena: ¿Un tercio?

Carmela: Oh, we learned this in school, 1.3.

Elena: Nine percent.

Carmela: How did you get that?

Elena: Cause you divide three times the decimal.

Selene: What do you know about a third? Is it more than a half?

Elena: It's more.

Juan: It's less.

When Elena said, "More than half," we realized that she was unclear about more than the fraction-percent conversion. Because others were struggling too, we opted to slow down, to play with different ways to show these fractions. We treated this opportunity with great respect, conscious that it takes courage to learn the basics in public.

> Selene: If this is a complete bottle of 100 percent, half would be 50 percent and a third would be a little less ...

because it is three parts that make it complete and so it has to be of three equal parts, one, two, three.

We tore open a 24-pack of bottled water brought as a prop, inviting the group to show us one half. They quickly separated the 24-pack into two groups: 12 in each. We wrote "12 is half of 24" on the board: "Two groups of 12 to make 24." We asked them to explain how one half is like 12/24. The school-age participants smiled knowingly. We all agreed that another way to say one half is one out of every two.

"One third of the bottled water sold in the United States is from the tap." We challenged the group to show us with the 24-pack. Now they could see that "one out of two" is more than "one out of three." Half the 24-pack was 12; one third was only eight. A participant demonstrated

Martha Merson

one fourth to emphasize the relationship between the change in the denominator (increase) and the number of bottles set apart (decrease).

We spent a little time on the second statistic, especially because Sandra had mentioned that bottled water costs double. Her statement communicated the gist of the situation: Consumers are paying dearly for water, but it massively understated the amount. The film draws an analogy between paying for bottled water and a consumer paying $10,000 for a hamburger. To explore the comparison between the cost of bottled water and tap water, the group listed current prices and calculated price tags 2,000 times higher. For one gallon of gasoline—$8,000!

Selene summarized, "Would one spend that much money on [gas]? But that is what we are spending in reality on water."

Murmurs and head shaking indicated the group members were impressed. Bottled water is overpriced, and bottled water companies are generating big bucks from consumers' willingness to pay the price.

## The Transparency of Water: See for Yourself

Ultimately, we want residents to make informed decisions based on data rather than on suspicion or misinformation or even faith in us. In the beginning of the video, Leonard reminds viewers that "in many ways bottled water is less regulated than tap." The Food and Drug Administration (FDA), which regulates bottled water in the United States, "does not have the ability to

require the submission to the agency of results from the testing conducted by and on behalf of bottled water manufacturers, and that FDA does not have specific authority to mandate the use of certified laboratories," according to a testimony given by the FDA to the U.S. Department of Health and Human Services.[3]

We felt that it was vital to draw this distinction because although there might be legitimate concerns of what lurks in tap water, at least it is a matter of public record. People voice all kinds of fears about the tap water, but they can read water test results, answer some of their own questions, and make specific concerns and demands of their government. Having information about our drinking water allows people to take action. This is the case in California where residents are fighting to be the first state to set a limit on the carcinogen hexavalent chromium.[4]

To this end, we examined Chicago's water report (see sample below), which we distributed to participants. Municipalities have to report to

**COMPREHENSIVE CHEMICAL ANALYSIS**
Page 1 of 2
**CITY OF CHICAGO - DEPARTMENT OF WATER MANAGEMENT - BUREAU OF WATER SUPPLY**
**WATER QUALITY DIVISION-WATER PURIFICATION LABORATORIES**
LABORATORY ACCREDITATION NUMBER: 100228

SAMPLE COLLECTION DATE: March 12, 2009 - 1st Quarter

| PARAMETER | IEPA MCL | DETERMINED AS | STORET NUMBER | SOUTH WATER PURIFICATION PLANT | | | | JARDINE WATER PURIFICATION PLANT | | | | |
| --- | --- | --- | --- | --- | --- | --- | --- | --- | --- | --- | --- | --- |
| | | | | RAW LAKE | OUTLETS 73rd Street | OUTLETS 79th Street | ***DISTRIBUTION SOUTH | RAW LAKE | OUTLETS North | OUTLETS Central | ***DISTRIBUTION Central | ***DISTRIBUTION North |
| TEMPERATURE | | °C | 00010 | 4 | 3 | 3 | 5 | 4 | 8 | 7 | 4 | 4 |
| TURBIDITY | 0.5 | N.T.U. | 82079 | 6.4 | 0.10 | 0.10 | 0.10 | 4.3 | 0.10 | 0.10 | 0.10 | 0.10 |
| THRESHOLD ODOR, STRAIGHT | *3 | T.O.N | 00086 | 1 E | 1 Cc | 1 Cc | 2 Cc | 1 E | 1 Cc | 1 Cc | 1 E | 1 E |
| THRESHOLD ODOR, DECHLORINATED | *3 | T.O.N. | | 1 E | 1 M | 1 M | 2 M | 1 E | 1 M | 1 M | 1 E | 1 E |
| COLOR | *15 | Pt.-Co. Units | 00080 | 0 | 0 | 0 | 0 | 0 | 0 | 0 | 0 | 0 |
| pH | *6.5-8.5 | STD. Units | 00040 | 8.17 | 7.87 | 7.86 | 7.60 | 8.27 | 7.93 | 7.92 | 7.68 | 7.69 |
| FREE CHLORINE RESIDUAL | | CL₂, mg/L | 50064 | -- | 1.00 | 1.00 | 0.69 | -- | 1.01 | 1.00 | 0.74 | 0.70 |
| SATURATION INDEX, LANGELIER | | UNITS +/- | | 0.29 | -0.07 | -0.05 | -0.28 | 0.35 | 0.00 | 0.00 | -0.29 | -0.24 |
| ALKALINITY, PHENOLPHTHALEIN | | CaCO3, mg/L | 00415 | 0 | 0 | 0 | 0 | 0 | 0 | 0 | 0 | 0 |
| ALKALINITY, TOTAL | | CaCO3, mg/L | 00410 | 118 | 103 | 103 | 102 | 115 | 104 | 104 | 103 | 103 |
| BROMIDE | | Br, mg/L | 71870 | ND | ND | ND | ND | ND | ND | ND | ND | ND |
| CHLORIDE | *250 | Cl, mg/L | 00940 | 16.0 | 15.0 | 15.2 | 14.6 | 17.3 | 16.0 | 15.7 | 14.3 | 14.4 |
| FLUORIDE | 4 | F, mg/L | 00951 | 0.13 | 1.04 | 1.03 | 0.97 | 0.01 | 0.91 | 0.90 | 0.90 | 0.91 |
| SULFATE | *250 | SO4, mg/L | 00945 | 24.7 | 29.9 | 28.9 | 28.8 | 24.9 | 28.5 | 28.5 | 28.0 | 27.9 |
| HARDNESS | | CaCO3, mg/L | 00900 | 144 | 141 | 141 | 142 | 142 | 138 | 137 | 137 | 135 |
| CALCIUM | | Ca, mg/L | 00916 | 31.4 | 30.6 | 32.7 | 30.3 | 34.4 | 32.7 | 33.9 | 28.9 | 31.6 |
| MAGNESIUM | | Mg, mg/L | 00927 | 11.7 | 11.4 | 11.7 | 11.9 | 12.8 | 12.4 | 12.3 | 11.0 | 11.6 |
| POTASSIUM | | K, mg/L | 00937 | 1.1 | 1.4 | 1.4 | 1.1 | 1.5 | 1.3 | 1.3 | 1.5 | 1.5 |
| SODIUM | | Na, mg/L | 00006 | 9.2 | 8.2 | 8.5 | 7.9 | 10.0 | 8.8 | 8.7 | 7.3 | 7.7 |
| SOLIDS, TOTAL DISSOLVED | *500 | TDS, mg/L | 00150 | 135 | 137 | 177 | 129 | 170 | 167 | 167 | 164 | 156 |
| SOLIDS, TOTAL | | Tot. Sol., mg/L | 00500 | 191 | 185 | 186 | 185 | 190 | 191 | 191 | 179 | 183 |
| TOTAL ORGANIC CARBON | | NPOC, mg/L | 00680 | 1.62 | 1.31 | 1.32 | 1.32 | 1.73 | 1.40 | 1.35 | 1.32 | 1.32 |
| OXYGEN DEMAND, CHEMICAL | | O, mg/L | 00335 | 6.5 | <5 | <5 | <5 | <5 | <5 | <5 | <5 | <5 |
| NITROGEN, AMMONIA | | N, mg/L | 00610 | 0.05 | 0.03 | <0.03 | 0.03 | 0.04 | <0.03 | <0.03 | <0.03 | 0.03 |
| NITROGEN, NITRATE | 10 | N, mg/L | 00620 | 0.500 | 0.420 | 0.329 | 0.328 | 0.562 | 0.349 | 0.346 | 0.326 | 0.325 |
| NITROGEN, NITRITE | 1 | N, mg/L | 00615 | <0.02 | <0.02 | <0.02 | <0.02 | <0.02 | <0.02 | <0.02 | <0.02 | <0.02 |
| NITROGEN, TOTAL KJELDAHL | | N, mg/L | 00625 | O/S | O/S | O/S | O/S | O/S | O/S | O/S | O/S | O/S |
| ORTHOPHOSPHATE | | PO4, mg/L | 00660 | <0.05 | 0.530 | 0.536 | 0.547 | <0.05 | 0.426 | 0.426 | 0.439 | 0.426 |
| PHOSPHATE, TOTAL | | PO4, mg/L | 00650 | <0.05 | 1.166 | 1.198 | 1.164 | <0.05 | 0.958 | 0.966 | 0.888 | 0.912 |
| CYANIDE, TOTAL | 200 | CN, ug/L | 00720 | <5 | <5 | <5 | <5 | <5 | <5 | <5 | <5 | <5 |
| RADIOACTIVITY, GROSS ALPHA | 15 | pCi/L | 01501 | <1 | <1 | <1 | <1 | <1 | <1 | <1 | <1 | <1 |
| RADIOACTIVITY, GROSS BETA | 50 | pCi/L | 03501 | 4 | 2 | 3 | 4 | 2 | 2 | 3 | 2 | 2 |

* Federal/State Secondary MCL's    ** Action Level    ***Distribution samples are composited.    ND - non detected

residents annually, but few people know that or have ever read them. We asked participants: "What do you know? What do you want to know?" When Norma admitted feeling overwhelmed, Selene explained:

> Selene: Parameters, what they tested for, go down the side. See anything familiar?
> All: Chlorine, fluoride, copper, lead.
> Juan: The water has all of this?
> Carmela: *What!* There's cyanide in our water?
> Selene: We'll see. These are *all* the things they test for. In the next column, we have numbers from the Illinois Environmental Protection Agency. What do you think? Why would they put numbers in a water chart?
> Elena: Percentages?
> Selene: For what purpose?
> Yoana: To communicate?
> Selene: Yes, they are communicating that more than 250 micrograms of cyanide is unsafe, because 250 is the maximum contaminant level or MCL. If we see higher, it's for sure a known risk. What else?
> Elena: Some are blank.
> Sandra: Because they are trying to figure out if it is good or bad?

The federal guidelines for "Maximum Contaminant Levels" in drinking water regulate about 100 of the chemicals in use. Municipalities test for other chemicals, but under the Illinois EPA regulations, there is no set limit for comparison so the cell is blank.

Carmela voiced a question about the abbreviation "ND." Quickly we set up a demonstration of "Not Detected," with a postage scale.

> Martha: Lab equipment tests for how much contaminant is present. Will every scale give an exact reading? For example, this scale, is it accurate for small objects?
> Elena: Try this pen cap.
> Maria: Try this paper clip.

The scale registered nothing until we tried a book. Everyone witnessed reporting limits firsthand: a zero reading does not mean no amount is present. Below a certain amount, a scale simply won't register. ND means any amount of contaminant from zero to the reporting limit could be present. Anyone reading environmental quality reports should check that the reporting limit is set below the amount considered safe.

Next we pointed out the columns listing the level of contaminants in the "raw" lake water. After treatment, sodium levels go down, while chlorine levels increase. Pairs picked one parameter, tracing how the level changed before and after treatment.

> Carmela: Cyanide is allowed in drinking water up to 200 micrograms per liter and it shows that it came in at ... wait, when there is a "less than" sign ... what does that mean?
> Selene draws "<" on the board.
> Carmela: The alligator eats the bigger number! (Everyone laughs.) So less than five.
> Selene: Exactly. That's how I remember ... I used to draw little teeth. What the alligator is eating tells us that the actual amount is less than whatever number it's eating.

The participants nodded and Carmela finished her assessment: The amounts for cyanide before and after treatment were less than five.

Less than five what? The column specifying units indicates less than five micrograms per liter.

The time went so quickly. We had to put off an exploration of units as well as the critical notion of "safe" levels. Identifying (or explaining or determining) "safe" levels is a frustrating endeavor. For carcinogens such as arsenic, there is no such thing as a safe level, only what has been deemed "acceptable" risk. Generally standards are set to protect safety and are set based on known risks, yet many agree that standards are inadequate, because researchers know little about synergistic effects, the likely accumulated health impacts from exposure to a variety of chemicals across a lifespan. Over time, researchers may compile a body of evidence that shows negative health effects at lower levels than previously thought, but it can take years for regulations to catch up. Further, setting standards is an imperfect process, one that historically takes into account costs to business and government as well as research findings. Complex as it is to make sense of levels and standards, clean water activists believe politicians will succumb to pressure from industry (as they have on efforts to tighten clean air regulations and arsenic standards) if communities are silent on these issues.

## Participants' Responses and Our Reflections

With limited time for explorations, it is tempting to avoid messy data sets and questions that have no easy answers (such as "Is <5 μg/l of cyanide safe?"). Yet we wonder whose purposes are served if we take a pass. SfA (free online) resources include activities and data sets that make teaching the math of environmental data a bit easier. In addition, SfA's "Smart Moves" offers nine hints for facilitators and participants to take control of the math (see sfa.terc.edu). The Smart Moves invite individuals at all math levels to interpret and communicate environmental data. Although the Common Core State Standards set forth a prescribed sequence for math learning, our experience shows that local, relevant data spark engagement at all

levels. Regardless of their past success or failure with math, participants grasped concepts like reporting limits, explored persistent misunderstandings like the relative size of a half and a third, coordinated information from rows and columns to identify contaminant levels, and young participants made connections to school learning.

A New England-based activist, Jackie Elliott, once told us: "Proposers or developers of a nasty product go to communities that are naive, have low levels of education, are politically powerless, and are compromised with economic stress."

Elliott maintained that with a handful of people who have done their homework and go to the hearings armed with numbers, the proposers realize they can't put on their usual dog and pony show.

We considered this workshop a success. As math educators, we were pleased that the people stayed with the math instead of counting themselves out. The group got started on the path toward digging into the numbers. As environmentalists we observed participants make connections

steve@greenberg-art.com VENTURA COUNTY STAR '08 GREENBERG

EVEN THOUGH WE HAVE SOME OF THE BEST AND SAFEST TAP WATER IN THE WORLD...

AMERICANS STILL BUY UPWARD OF 28 BILLION BOTTLES OF WATER A YEAR.

MANUFACTURING THIS USES AS MUCH AS 45 TO 50 MILLION BARRELS OF OIL.

80 PERCENT OF THE USED BOTTLES WIND UP IN LAND-FILLS, NOT IN RECYCLING.

PLASTIC BOTTLES HEATED IN THE SUN CAN LEACH CHEMICALS INTO THE WATER.

AND ON A PER-OUNCE BASIS, BOTTLED WATER COSTS TWICE OR MORE WHAT GASOLINE DOES, AND CAN BE UP TO 1,000 TIMES THE COST OF TAP WATER.

BUT ISN'T IT WORTH IT, SINCE THEY NO DOUBT BOTTLE IT FROM SOME PRISTINE, MYSTERIOUS SOURCE?

Status Water

between purchasing choices and environmental consequences like mountains and mountains of trash. As organizers we saw workshop participants acquiring strategies to restate facts to support their beliefs. That one third of bottled water comes from the tap confirmed Elena's sense: "I always felt that tap water was as good. ... I just found it hard to believe that that [bottled] water was better than tap water."

On a short evaluation participants told us what they valued learning:

I can drink water from the sink.
I'll know not to buy bottled water and save money.
Cost of H2O (bottled) 2,000 x more!

These comments speak to both individuals' choices and the larger economic issues.

And the conversation continued as people left:

Yoana: They [the advertisements] are trying to confuse the people; people should have access to water without paying so much.

Months later, the participants of this workshop joined a protest of the mayor's move to privatize Chicago's water system. They had the background to contribute to discourse on who is in the best position to monitor and safely deliver drinking water.

## Conclusion

Clean drinking water supply is in a precarious position, threatened by fracking, pesticide runoff, and, in Chicago, privatization. Even in cities where the water routinely gets high marks, residents can't afford to be complacent. Access to clean water, as Annie Leonard says, is our birthright, but in a capitalist economy, nearly everything is for sale and the water (system) we all rely on and need to live needs vigilant protection.

Many of us memorize and spout statistics that affirm our beliefs without so much as a glance at data sets like water quality reports. As educators and organizers we advocate interpreting data to describe conditions. We advocate shedding light on how to calculate, quantify, and explain the costs of injustice. Workshops like this one set the stage for broader involvement in developing and understanding these statistics. This is part of a respectful approach to convincing people who've grown accustomed to buying water that they are shouldering an unjust financial and health burden. Intuitively many know the situation is inequitable, but facing the data can ignite a sense of urgency. Then the hard work of identifying steps for collective action begins. □

## ENDNOTES

The leadership sessions were based on a curriculum called *People United for Dignity, Democracy, and Justice.*

1. March H. Gorelick, MD, MSCE. (2011, June 6). "Perceptions About Water and Increased Use of Bottled Water in Minority Children." Archives of *Pediatrics & Adolescent Medicine.* Vol 165, No. 10. Retrieved November 1, 2012 from http://archpedi.jamanetwork.com/article.aspx?articleid=1107603#AuthorInformation

2. EWG (Environmental Working Group). 2008. Bottled Water Quality Investigation: 10 Major Brands, 38 Pollutants. Retrieved November 1, 2012 from: http://www.ewg.org/reports/BottledWater/Bottled-Water-Quality-Investigation

3. Joshua M. Sharfstein, MD (July 8, 2009). Testimony on Regulation of Bottled Water. *HHS.GOV: US Department of Health and Human Services.* Retrieved November 29, 2012 from http://www.hhs.gov/asl/testify/2009/07/t20090708a.html

4. Lyndsey Layton. (December 19, 2010). "Probable Carcinogen Hexavalent Chromium Found in Drinking Water of 31 U.S. cities." *The Washington Post.* Retrieved November 29, 2012 from http://www.washingtonpost.com/wp-dyn/content/article/2010/12/18/AR2010121802810.html

# CHAPTER TWENTY-NINE

# Write the Truth: Presidents and Slaves

## BY BOB PETERSON

During a lesson about George Washington and the American Revolution, I explained to my 5th graders that Washington owned 317 slaves. One student added that Thomas Jefferson also was a slave owner. And then, in part to be funny and in part expressing anger—over vote fraud involving African Americans in the then-recent 2000 election and the U.S. Supreme Court's subsequent delivery of the presidency to George W. Bush—one of my students shouted: "Bush is a slave owner, too!"

"No, Bush doesn't own slaves," I calmly explained. "Slavery was finally ended in this country in 1865."

Short exchanges such as this often pass quickly and we move onto another topic. But this time one student asked: "Well, which presidents were slave owners?"

She had me stumped. "That's a good question," I said. "I don't know."

Thus began a combined social studies, math, and language arts project in which I learned along with my students and which culminated in a fascinating exchange between my students and the publishers of their U.S. history textbook.

After I admitted that I had no clue exactly how many presidents owned slaves, I threw the challenge back to the students. "How can we find out?" I asked.

"Look in a history book," said one. "Check the internet," added another.

I realized that I had entered one of those "teachable moments" when students show genuine interest in exploring a particular topic. Yet I had few materials about presidents and slaves and no immediate idea of how to engage 25 students on the subject.

I also recognized that this was a great opportunity to create my own curriculum, which might help students look critically at texts while encouraging their active participation in doing meaningful research. Such an approach stands in sharp contrast to the "memorize the presidents" instruction that I suffered through growing up and that too many students probably still endure. I seized the opportunity.

**Action Research by Students**

First, I had a student write the question "Which presidents were slave owners?" in our class notebook, "Questions We Have." I then suggested that a few students form an "action research group," which in my classroom means an ad hoc group of interested students researching a topic and then doing something with what they learn. I asked for volunteers willing to work during recess. Several boys raised their hands, surprising me because I would have guessed that some of them would have much preferred going outside to staying indoors conducting research.

At recess time, Raul and Edwin were immediately in my face. "When are we going to start the action research on the slave presidents?" they demanded. I told them to look in the back of our school dictionaries for a list of U.S. presidents while I got out some large construction paper.

The dictionaries, like our social studies text, had little pictures of each president with

## The student researchers were excited to present their findings and decided to do so as part of a math class.

some basic information. "Why doesn't it show Clinton?" Edwin commented. "He's been president forever."

I think, yeah, Clinton's been president four fifths of this 10-year-old's life. But I kept that thought to myself and instead replied, "The book is old."

"Why don't they just tell whether they have slaves here in this list of presidents?" asked Edwin. "They tell other things about presidents."

"Good question," I said. "Why do you think they don't tell?"

"I don't know, probably because they don't know themselves."

"Maybe so," I responded. "Here's what I'd like you to do. Since slavery was abolished when Lincoln was president, and since he was the 16th president, draw 16 lines equal distance from each other and list all the presidents from Washington to Lincoln, and then make a yes-or-no column, so we can check off whether they owned slaves." (See page 234 for a sample chart.)

I was soon to find out that filling in those columns was easier said than done.

When my students and I began investigating which presidents owned slaves, our attempts focused on traditional history textbooks and student-friendly websites from the White House and the Smithsonian Institution. These efforts turned up virtually nothing.

We then pursued two different sources of information: history books written for adults and more in-depth websites.

I brought in two books that were somewhat helpful: James Loewen's *Lies My Teacher Told Me* and Kenneth O'Reilly's *Nixon's Piano: Presidents and Racial Politics from Washington to Clinton.* By using the indexes and reading the text out loud, we uncovered facts about some of the presidents.

We also used the web search engines Google and AltaVista and searched on the words "presidents" and "slavery." We soon learned we had to be more specific and include the president's name and "slavery"—for example, "President George Washington" and "slavery." Some results were student-friendly, such as the mention of Washington's slaves (and some of their escapes) at www. mountvernon.org. There was also a bill of sale for a slave signed by Dolley Madison, the wife of president James Madison (for a link to the document, see www.rethinkingschools.org/math).

Many websites had a large amount of text and were beyond the reading level of many of my students. So I cut and pasted long articles into word-processing documents so we could search for the word "slave" to see if there was any specific mention of slave ownership.

In their research, students often asked, "How do we know this is true? Our history books aren't telling the truth, why should we think this does?"

I explained the difference between primary and secondary sources and how a primary source—like a bill of sale or original list of slaves—was pretty solid evidence. To help ensure accuracy, the students decided that if we used secondary sources, we needed to find at least two different citations.

### Bits and Pieces of Information

In the next several days the students, with my help, looked at various sources. We checked our school's children's books about presidents, our social studies textbook, a 1975 World Book Encyclopedia, and a CD-ROM encyclopedia. We found nothing about presidents as slave owners.

I had a hunch about which presidents owned slaves, based on what I knew in general about the presidents, but I wanted "proof" before we put a check in the "yes" box. And though my students wanted to add a third column—explaining how many slaves each slave-owning president had—that proved impossible. Even when we did find information about which presidents owned slaves, the numbers changed depending on how many slaves had been bought, sold, born, or died.

In our research, most of the information dealt with presidential attitudes and policies

## EIGHT COMPANIES EARN MORE THAN HALF THE WORLD'S POPULATION!

### ONE-DOLLAR BILL GRAPHIC TEACHING SUGGESTIONS

Have students research data on the distribution of the world's wealth to see why Polyp, a graphic artist, created the image on the page at right.

Using such data, have students draw

graphs and cartoons to demonstrate the inequality in the world's wealth.

Also do the simulation "Poverty and World Wealth," page 89.

toward slavery. It was difficult to find specific information on which presidents owned slaves. To help the investigation, I checked out a few books for them from our local university library.

Overall, our best resource was the internet. The best sites required adult help to find and evaluate, and I became so engrossed in the project that I spent a considerable amount of time at home surfing the web. The "student-friendly" websites with information about presidents—such as the

know themselves." (Given more time, we might have explored this matter further, looking at who produces textbooks and why they might not include information about presidents' attitudes about racism and slavery.)

During our research, my students and I found bits and pieces of information about presidents and slavery. But we never found that one magic

resource, be it book or website, that had the information readily available. Ultimately, though, we came up with credible data.

White House's gallery of presidents, www.whitehouse.gov/about/presidents—don't mention that Washington enslaved African Americans or that Jefferson did. Other popular sites with the same glaring lack of information are the Smithsonian Institution, www.si.edu, and the National Museum of American History, www.american-history.si.edu/presidency.

As we did the research, I regularly asked, "Why do you think this doesn't mention that the president owned slaves?" Students' responses varied including, "They're stupid," "They don't want us kids to know the truth," "They think we're too young to know," and "They don't

I'm a history buff, and I had thought I was on top of the question of presidents and slavery. I was quite amazed, and didn't hide my amazement from our action research team, when they discovered that two presidents who served after Lincoln—Andrew Johnson and Ulysses S. Grant—had been slave owners. While the students taped an extension on their chart, I explained that I was not totally surprised about Johnson because he had been a Southerner. But it was a shock that Grant had owned slaves. "He was the commander of the Union army in the Civil War," I explained. "When I first learned about the Civil War in elementary school, Grant and Lincoln were portrayed as saviors of the Union and freers of slaves."

## Slave-Owning Presidents

When I told the entire class how Grant's slave-owning past had surprised me, Tanya, an African American student, raised her hand and said, "That's nothing. Lincoln was a slave-owner, too."

I asked for her source of information and she said she had heard that Lincoln didn't like blacks. I thanked her for raising the point, and told the class that while it was commonly accepted by historians that Lincoln was not a slave-owner, his attitudes toward blacks and slavery were a source of much debate. I noted that just because a president didn't own slaves didn't mean that he supported freedom for slaves or equal treatment for people of different races.

I went into a bit of detail on Lincoln, in part to counter the all-too-common simplification that Lincoln unequivocally opposed slavery and supported freedom for blacks. I explained that while it's commonly believed that Lincoln freed enslaved Americans when he signed the Emancipation Proclamation, the document actually frees slaves only in states and regions under rebellion—it did not free slaves in any of the slaveholding states and regions that remained in the Union. In other words, Lincoln "freed" slaves everywhere he had no authority and withheld freedom everywhere he did. Earlier, in Lincoln's first inaugural address in March 1861, he promised slaveholders that he would support a constitutional amendment forever protecting slavery in the states where it then existed—if those states would only remain in the Union.

By the time we finished our research, the students had found that 10 of the first 18 presidents were slave owners: George Washington,

> "[W]hile the word 'racism' does not appear, the subject of unfair treatment of people because of their race is addressed on page 467."
>
> — The vice president of Harcourt School Publishers

Thomas Jefferson, James Madison, James Monroe, Andrew Jackson, John Tyler, James K. Polk, Zachary Taylor, Andrew Johnson, and Ulysses S. Grant.

Those that didn't own slaves were: John Adams, John Quincy Adams, Martin Van Buren, William Harrison, Millard Fillmore, Franklin Pierce, James Buchanan, and, despite Tanya's assertion, Abraham Lincoln.

The student researchers were excited to present their findings to their classmates and decided to do so as part of a math class. I made blank charts for each student in the class, and they filled in information provided by the action research team: the names of presidents, the dates of their years in office, the total number of years in office, and whether they had owned slaves. Our chart started with George Washington, who assumed office in 1789, and ended in 1877 when the last president who had owned slaves, Ulysses Grant, had left office.

We then used the data to discuss this topic of presidents and slave-owning within the structure of ongoing math topics in my class: "What do the data tell us?" and "How can we construct new knowledge with the data?"

Students, for example, added up the total number of years in which the United States had a slave-owning president in office, and compared that total to the number of years in which there were non-slave-owning presidents in office. We figured out that in 69 percent of the years between 1789 and 1877, the United States had a president who had been a slave-owner.

One student observed that only slave-owning presidents served more than one term. "Why didn't they let presidents who didn't own slaves serve two terms?" another student pondered.

Using the data, the students made bar graphs and circle graphs to display the information. When they wrote written reflections on the math lesson, they connected math to content. One boy wrote: "I learned to convert fractions to percent so I know that 10/18 is the same as 55.5 percent. That's how many of the first 18 presidents owned slaves." Another girl observed, "I learned how to make pie charts and that so many more presidents owned slaves than presidents who didn't own slaves."

During a subsequent social studies lesson, the three students who had done most of the research explained their frustrations in getting information. "They hardly ever want to mention it [slaves owned by presidents]," explained one student. "We had to search and search."

## Mini-Unit Objectives

Specific objectives for this mini-unit, such as reviewing the use of percentages, emerged as the lessons themselves unfolded. But its main purpose was to help students critically examine the actions of early leaders of the United States and become skeptical of textbooks and government websites as sources that present the entire picture. I figure that if kids start questioning the "official story" early on, they will be more open to alternative viewpoints later on. While discovering which presidents were slave owners is not an in-depth analysis, it pokes an important hole in the godlike mystique that surrounds the "founding fathers." If students learn how to be critical of the icons of the American past, hopefully it will give them permission and tools to be critical of the elites of America today.

Besides uncovering some hard-to-find and uncomfortable historical truths, I also wanted to encourage my students to think about why these facts were so hard to find and to develop a healthy skepticism of official sources of information. I showed them two quotations about Thomas Jefferson. One was from a recently published, 5th-grade history textbook, *United States: Adventures in Time and Place*

> "As far as the laws of mathematics refer to reality, they are not certain, and as far as they are certain, they do not refer to reality."
>
> **— Albert Einstein,**
> *Sidelights on Relativity,* 1923

(New York: Macmillan/McGraw-Hill, 1998) which read: "Jefferson owned several slaves in his lifetime and lived in a slave-owning colony. Yet he often spoke out against slavery. 'Nothing is more certainly written in the book of fate than that these people are to be free.'" (page 314)

The other quotation was from James Loewen's *Lies My Teacher Told Me.* Loewen writes:

> Textbooks stress that Jefferson was a humane master, privately tormented by slavery and opposed to its expansion, not the type to destroy families by selling slaves. In truth, by 1820 Jefferson had become an ardent advocate of the expansion of slavery to the western territories. And he never let his ambivalence about slavery affect his private life. Jefferson was an average master who had his slaves whipped and sold into the Deep South as examples to induce other slaves to obey. By 1822, Jefferson owned 267 slaves. During his long life, of hundreds of different slaves he owned, he freed only three and five more at his death—all blood relatives of his. (page 140)

We talked about the different perspective each quote had toward Jefferson and toward what students should learn. My students' attention immediately turned to the set of spanking new history textbooks that had been delivered to our classroom that year as part of the district-wide social studies adoption. Some students assumed that our new textbook *United States* (Harcourt Brace, 2000) was equally as bad as the one I quoted from. One student suggested we just throw the books away. But I quickly pointed out they were expensive, and that we could learn from them even if they had problems and omissions.

I then explained what an omission was and suggested that we become "textbook detectives" and investigate what our new social studies text said about Jefferson and slavery. I reviewed how to use an index and divided all page references for Jefferson among small groups of students. The groups read the pages, noted any references to Jefferson owning slaves, and then reported back to the class. Not one group found a single reference.

Not surprisingly, the students were angry when they realized how the text omitted such important information. "They should tell the truth!" one student fumed.

## No Mention of Racism

I wanted students to see that the textbook's omissions were not an anomaly, but part of a pattern of ignoring racism in the United States—in the past and in the present.

In the next lesson, I started by writing the word "racism" on the board. I asked the kids to look up "racism" in the index of their social studies book. Nothing. "Racial discrimination." Nothing.

"Our school should get a different book," one student suggested.

"Good idea," I said, "but it's not so easy." I told my students that I had served on a committee that had looked at the major textbooks published for 5th graders and that none of them had dealt with racism or slavery and presidents.

Students had a variety of responses:

"Let's throw them out."

"Let's use the internet."

"Write a letter to the people who did the books."

I focused in on the letter-writing suggestion and reminded them that before we did so, we had to be certain that our criticisms were correct. The students then agreed that in small groups they would use the textbook's index and read what was said about all the first 18 presidents, just as we had done previously with Jefferson.

None of the groups found any mention of a president owning a slave.

## Letters as Critique and Action

In subsequent days, some students wrote letters to the textbook publisher. Michelle, a white girl, was particularly detailed. She wrote: "I am 11 years old and I like to read and write. When I am reading I notice every little word and in your social studies book I realize that the word "racism" is not in your book. You're acting like it is a bad word for those kids who read it." She went on to criticize the book for not mentioning that any presidents had slaves: "I see that you do not mention that some of the presidents had slaves. But some of them did. Like George Washington had 317 slaves. So did Thomas Jefferson. He had 267 slaves." She continued: "If you want to teach children the truth, then you should write the truth." (Michelle's letter and some of the student-made charts were also printed in our school newspaper.)

We mailed off the letters, and moved on to new lessons. Weeks passed with no response and eventually the students stopped asking if the publishers had written back. Then one day a fancy-looking envelope appeared in my mailbox addressed to Michelle Williams. She excitedly opened the letter and read it to the class.

Harcourt School Publishers Vice President Donald Lankiewicz had responded to Michelle at length. He wrote that "while the word 'racism' does not appear, the subject of unfair treatment of people because of their race is addressed on page 467." He also argued: "There are many facts about the presidents that are not included in the text simply because we do not have room for them all."

Michelle wrote back to Lankiewicz, thanking him but expressing disappointment. "In a history book you shouldn't have to wait till page 467 to learn about unfair treatment," she wrote. As to his claim that there wasn't room for all the facts about the presidents, Michelle responded: "Adding more pages is good for the kids because they should know the right things from the wrong. It is not like you are limited to certain amount of pages. ... All I ask you is that you write the word "racism" in the book and add some more pages in the book so you can put most of the truth about the presidents."

Michelle never received a reply.

## Improving the Lesson

Michelle and the other students left 5th grade soon after the letter exchange. In the flurry of end-of-year activities, I didn't take as much time to process the project as I might have. Nor did I adequately explore with students the fact that most non-slave-owning presidents exhibited pro-slavery attitudes and promoted pro-slavery policies.

But the larger issue, which critical teachers struggle to address, is why textbook publishers and schools in general do such a poor job of helping students make sense of the difficult issues of race. We do students a disservice when we sanitize history and sweep uncomfortable truths under the rug. We leave them less prepared to deal with the difficult issues they will face in their personal, political, and social lives. Granted, these are extremely complicated issues that don't have a single correct response. But it's important to begin with a respect for the truth and for the capacity of people of all ages to expand their understanding of the past and the present, and to open their hearts and minds to an ever-broadening concept of social justice.

I believe my students learned a lot from their research on presidents and slaves—and clearly know more than most Americans about which of the first 18 presidents owned slaves. I'm also hopeful they learned the importance of looking critically at all sources of information.

I know one student, Tanya, did. On the last day of school she came up to me amid the congratulatory good-byes and said, "I still think Lincoln owned slaves."

"You are a smart girl but you are wrong about that one," I responded.

"We'll see," she said, "You didn't know Grant had slaves when the school year started! Why should I always believe what my teachers say?" □

---

Some of the students' names in this article have been changed.

A chart for students to record which presidents owned slaves appears on page 234.

**Author's Note:** About two years after I completed the research on slave-owning presidents with my students, folks at Wesleyan University put up a wonderful website, UnderstandingPrejudice.org. The page www.understandingprejudice.org/slavery includes extensive information on presidents who owned slaves. I learned from this website that three presidents not on my list also owned slaves: Martin Van Buren, William Henry Harrison, and James Buchanan. I was grateful for the additional information on this website, which opens up all sorts of new teaching possibilities.

# PRESIDENTIAL FACTS

## WHICH PRESIDENTS OWNED SLAVES AND FOR HOW LONG?

| # | NAME OF PRESIDENT | YEARS IN OFFICE | TOTAL YEARS SERVED | OWNED SLAVES? | # OF SLAVES OWNED | YEARS SERVED BY SLAVE OWNER | YEARS SERVED BY NON-SLAVE OWNER |
|---|---|---|---|---|---|---|---|
| 1 | George Washington | 1789–1797 | 8 | Yes | | 8 | |
| 2 | | | | | | | |
| 3 | | | | | | | |
| 4 | | | | | | | |
| 5 | | | | | | | |
| 6 | | | | | | | |
| 7 | | | | | | | |
| 8 | | | | | | | |
| 9 | | | | | | | |
| 10 | | | | | | | |
| 11 | | | | | | | |
| 12 | | | | | | | |
| 13 | | | | | | | |
| 14 | | | | | | | |
| 15 | | | | | | | |
| 16 | | | | | | | |
| 17 | | | | | | | |
| 18 | | | | | | | |

# LIBRARIES, BOOKS, AND BIAS

## USING MATH TO RAISE ISSUES OF REPRESENTATION

BY BOB PETERSON

Classroom and school libraries offer many opportunities to apply math while raising issues of social justice and representation. For example: Who has access to how many books? Who is represented in those books?

Educational researcher Stephen Krashen argues that a key to improving student reading is access to large quantities of books. To have access to lots of books, schools need well-funded libraries staffed by qualified librarians. Others have argued that students should have access to books that reflect their own cultural heritage and those of many other cultures.

### DO THE MATH

In small groups or as a whole class, students can investigate the following issues using ratios, percentages, graphs, and basic computation.

1.   What is the ratio of certified librarians to students in your school, school district, state, and nation? Make graphic representations of the data and then reflect on why there are differences, what impact these differences have, and what might be done to improve the situation.

2.   What is the ratio of books in your school library to the number of students in your school? Contact other schools or use the internet to find similar data for other schools. Compare and contrast the data from urban, suburban, and rural school districts.

3.   Who is represented in the biographies in your school library? Count the number of biographies and autobiographies in your library. Calculate the percentage of books in your library that are in these categories. Make graphs showing the percentage of biographies that are about different groups of people, such as:

•   Women/men.
•   African Americans/Latinos/Asians/Native Americans/Arab Americans/European Americans.
•   People from outside of the United States/people from the United States.
•   Rich people/poor and working people.

4.   What do you think would be a "fair" representation in your school library of these various categories? What would have to be done to make sure there is an improvement in representation at your school?

In each of the projects described above, students could interview librarians, school district officials, and school board members regarding these questions. ∎

---

For more information on Stephen Krashen's work, visit www.sdkrashen.com or www.rethinkingschools.org/math.

# THE MATHEMATICS OF LOTTERIES

## GOING BEYOND PROBABILITY

BY LAURIE RUBEL

Many curricula point teachers and students toward using lotteries as a way to teach and learn probabilistic concepts. However, lotteries can be analyzed in sociopolitical context. This activity points students toward investigating critical questions like: Who typically plays the lottery? Who promotes lotteries? Who benefits from lottery revenues?

Studies show that people earning more than $100,000 annually are less likely to be pathological gamblers than people with lower incomes. Also, while the majority of gamblers are white, African Americans are disproportionately represented. Thus, poor people of color are, in general, most impacted by the mathematics of lotteries.

### PROCEDURE

Groups of students can select one of their state lottery games and do three things:

1.  Explain the game rules.

2.  Provide a probability distribution of all possible outcomes.

3.  Determine the expected value of the game for various jackpot amounts.

For example, the New York MegaMillions game is played by choosing five different integers from 1 to 52 as well as a MegaBall number, chosen from 1 to 52. The winning five numbers can be chosen in any of 52C5, or 2,598,960, ways. There are 135,145,920 different MegaMillions tickets, but only one winner.

## MEGAMILLIONS

| PRIZE | 5 WINNING NUMBERS | MEGABALL | NUMBER OF TICKETS | PRIZE (ticket cost: $1) |
|---|---|---|---|---|
| 1ST | Match 5 | Match | 1 | Jackpot |
| 2ND | Match 5 | No match | 51 | $175,000–$1 |
| 3RD | Match 4 | Match | 235 | $5,000–$1 |
| 4TH | Match 4 | No match | 11,985 | $150–$1 |
| 5TH | Match 3 | Match | 10,810 | $150–$1 |
| 6TH | Match 3 | No match | 162,150 | $10–$1 |
| 7TH | Match 2 | Match | 551,310 | $7–$1 |
| 8TH | Match 1 | Match | 891,825 | $3–$1 |
| 9TH | No match | Match | 1,533,939 | $2–$1 |
| NO PRIZE | No match | No match | 131,983,614 | Lose $1 |

TABLE 1

Secondary prizes range from $175,000 to $2 for matching fewer than all of the numbers. Second prize winners match all five numbers but not the MegaBall number. So 51 of the 135,145,920 tickets win a second prize. Table 1 (page 236) shows the prize categories and the number of winning tickets. This shows that 3,162,306 tickets win, although about half win only $1. However, a staggering 131,983,614 are losing tickets!

## QUESTIONS

The National Gambling Impact Study, commissioned by Congress in 1996, suggests that some state lotteries target poor communities for their advertising. Students can examine the realities in their communities:

1.     Where are the lottery retailers and where is the lottery advertising most prevalent?

2.     Who, in their communities, seems to be playing the lottery?

3.     Among those who play the lottery, how much do they spend each week?

## FURTHER ANALYSIS

Once students consider these aspects of the state lottery, they can learn who benefits from lottery revenues. In New York, in the 2003–2004 fiscal year, the lottery directed 33 percent of its revenues ($1.9 billion) to state funds earmarked for education. New York City received about $703 million that year, earmarked toward education. But New York City education needs are far greater than those lottery revenues, so the lottery gets used as a kind of fiscal shell—extra money isn't given to education from lottery revenues; the lottery revenues enable the state to actually budget less for education. Students can analyze local changes in education budgets alongside rising lottery revenues to examine this issue.

Students can also uncover who benefits from lottery revenues, other than state departments of education, such as the lottery administration, lottery retailers, and game contractors. For example, in 2004, New York paid nearly $62 million in fees to its gaming contractor and additional money was spent on marketing and retailer commissions. Scientific Games Corp., a publicly traded company, is a lottery contractor who reported revenues of $560.9 million(!!) in 2003. The price of stock in this company has risen about $8 per share in the last year alone. Thus, an investment in the company which makes the lottery tickets and machines is astoundingly more sound than playing the lottery itself.

By their very nature, lottery games provide rich contexts for mathematical analysis. But distilling the mathematics of the games from their sociopolitical context, as is typically done, is a missed opportunity to help students use that mathematics to think critically about their own worlds. ■

# CHAPTER THIRTY

# Rethinking and Connecting Algebra to Real-World Issues

## BY SANDRA KINGAN

Incorporating a social advocacy component in a mathematics class means connecting the mathematics learned to pressing real-world issues. It can include having students investigate protecting the environment, understanding and managing climate change, fighting poverty, standing up for dignity and human rights, protecting privacy in an age of increasing surveillance, or any number of things that involve people living together in a community supporting one another. Students may be more likely to learn math if they realize its value to themselves and society.

Take, for example, the concept of functions, which one can teach in various ways. One way is for teachers to introduce the concept and notation and then have students read the graph of a function and identify where it is increasing, decreasing, or constant. A good example of a function to illustrate this concept is the graph of the Standard and Poor Case-Shiller Home Price Index against time. The S&P Case-Shiller Home Price Index is calculated from data on repeat sales of single-family homes. Students can download the data (from www.standardandpoors.com) and use a spreadsheet or graphic calculator to produce the graph. The graph below illustrates the housing crash. There is no controversy over how it is calculated, at least none outside academic circles.

Another good example of a function is the graph of carbon dioxide in the atmosphere. In 1958, Charles David Keeling, a geochemist, was working on a project in Hawaii that examined the carbon content of rivers and thought to also measure atmospheric carbon dioxide levels. Previous measurements made in the preindustrial era showed that carbon dioxide levels in the atmosphere were about 280 parts per million (ppm).

The observatory (elevation 11,000 feet) on the island of Hawaii is one of the few places in the United States where the air is clean and particle-free. Keeling didn't expect to find any pattern, based on available research, but was surprised to record the same number when he measured carbon dioxide levels every afternoon (roughly 312 parts per million). Morning measurements were slightly elevated since plants absorb carbon dioxide in the night and release it in the day.

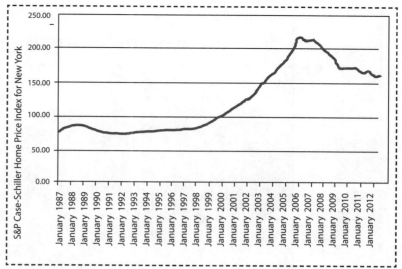

**FIGURE 1:** S&P CASE-SHILLER HOME PRICE INDEX FOR NEW YORK

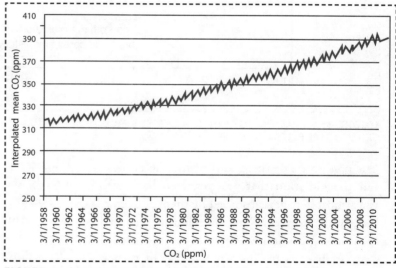

**FIGURE 2:** CARBON DIOXIDE LEVELS IN THE ATMOSPHERE

He also discovered that in the spring when plants grow, they absorb carbon dioxide and the levels dip, whereas in the fall, the decaying leaves return carbon dioxide to the atmosphere, and the levels rise. The Earth, in a sense, breathes. Most importantly, he discovered that carbon dioxide levels in the atmosphere are rising. The following graph obtained by plotting the carbon dioxide levels against time is called the Keeling Curve (data from noaa.gov). Observe that the carbon dioxide level in 2010 was more than 380 ppm. The general consensus among scientists is that 350 ppm is the safe upper limit for $CO_2$ in our atmosphere.

The atmosphere is made up of nitrogen (78.1 percent), oxygen (20.9 percent), and argon (0.93 percent), and less than 0.1 percent of the trace gases: carbon dioxide, water vapor, methane, nitrous oxide, ozone, and chlorofluorocarbons. The three main gases do not interact with the incoming solar radiation that heats the Earth. However, the trace gases play an important role in the Earth's energy budget. Sunlight

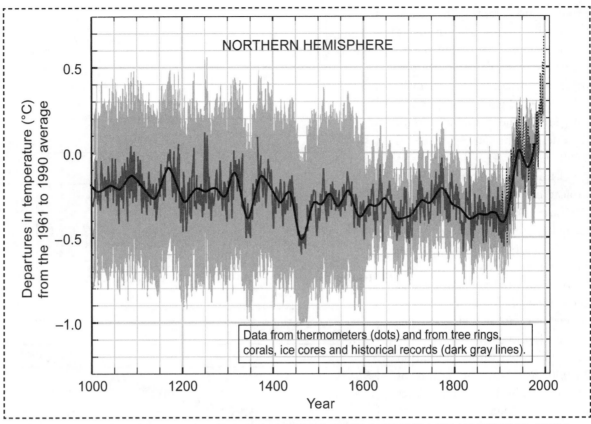

**FIGURE 3:** VARIATIONS OF THE EARTH'S SURFACE TEMPERATURE OVER THE LAST MILLENNIUM. WWW. GRIDA.NO/PUBLICATIONS/OTHER/IPCC_TAR

can pass through them and warm the surface, but the radiation that gets reflected back into the atmosphere gets absorbed by them and emitted to the surface. This mechanism (referred to as the greenhouse effect) keeps the surface comfortably warm, but an increase in carbon dioxide can cause a rise in surface temperature. This basic argument is accepted by all scientists.

In its 2001 report, the Intergovernmental Panel on Climate Change included a graph of reconstructed average global temperatures dating back a thousand years. This is known as the "hockey stick" graph due to its shape and shows that the average northern hemisphere temperature has risen steeply in the past 100 years. This graph generated much controversy with all sorts of politicians and religious groups getting in on the act.

Instead of alerting the public to the problem of rising carbon dioxide, the data became

a lightning rod for controversy. No one had directly and consistently measured the temperature before the 1800s, so scientists reconstructed earlier temperatures by measuring annual tree rings and isotopic ratios in corals, ice cores, and lake sediments—called "proxy data." Although no one disputed that temperatures had risen in the previous 150 years, some scientists questioned the statistical methods used to obtain one global average temperature for earlier years. They were concerned that there was theoretically no limit to the sources that could yield proxy data and wondered whether reconstructed temperature data from specific regions were representative of a global average temperature.

In 2006, the U.S. Congress asked the National Academy of Sciences to evaluate the validity of temperature reconstructions. Their report stated that although there may have been a few problems with the methods, subsequent

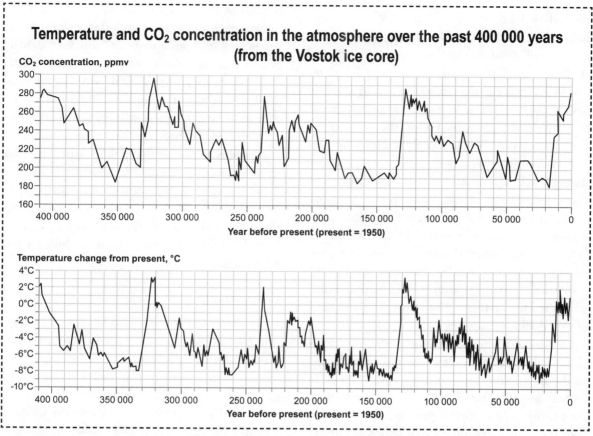

**FIGURE 4:** TEMPERATURE AND CARBON DIOXIDE CONCENTRATIONS OVER 400,000 YEARS. WWW.GRIDA.NO/PUBLICATIONS/VG/CLIMATE/PAGE/3057.ASPX

reconstructions based on improved scientific methods all reach the same conclusion that "warmth in the northern hemisphere was unprecedented during at least the last 1,000 years." There is no more controversy.

Scientists have also reconstructed temperature and atmospheric carbon dioxide levels using trapped gasses from polar ice cores. The graph above is difficult for some people to accept, especially those who dispute the age of the Earth and evolution. A 2008 Pew research study reported that of 35,556 people polled, 48 percent agreed with the statement: Evolution is the best explanation for the origins of human life on Earth. But 45 percent disagreed and 7 percent didn't know or refused to answer.

Perhaps there would be less disagreement if more people understood exponential functions and carbon dating. Exponential functions and their inverses, the logarithmic functions, are two of the most useful algebra topics in the curriculum with many interesting examples, most notably compound interest. Although compound interest is almost always covered (and with good reason), there may not be enough class time to cover carbon dating. It is worth making time for it as it could improve scientific literacy.

Radioactive dating is a technique used to date objects by comparing a naturally occurring radioactive isotope and its decay products. There are several techniques, one of which is carbon dating developed in 1949 by Willard Libby, for which he subsequently received the Nobel Prize in Chemistry. Carbon 14 and Carbon 12 are isotopes of Carbon. When organic matter dies, the Carbon 12 content remains fixed while the Carbon 14 (radioactive carbon) content decays with a half-life of 5,700 years. The initial ratio of Carbon 14 to Carbon 12 is $1:10^{12}$. An exponential decay model $y = C2^{kt}$ is used to estimate the age of dead organic material, where $y$ is the ratio of Carbon 14 to Carbon 12 present in $t$ years, $c$ is the initial ratio, and $k$ is the decay constant. The value of $k$ is calculated as $k = \frac{-1}{5700}$ using the fact that the half-life of Carbon 14 is 5,700 years. By taking the initial amount as $\frac{1}{10^{12}}$ we get $y = \frac{1}{10^{12}} 2^{\frac{-t}{5700}}$. Carbon dating works for objects that are no more than about 60,000 years old, with other methods for older objects. Scientists can fairly accurately measure the ratio of Carbon 14 to Carbon 12 ($y$ in the above equation) and can then compute the age. Any inaccuracies that arise are due to errors in instrumentation, not mathematics, which is why a range is given for the age, instead of one definitive number.

Teachers can incorporate the above applications in an algebra class, instead of outdated textbook applications, and they can serve as launching pads for further discussions, time permitting. These applications are not meant to take the place of standard math topics, but are meant to motivate students to study mathematics and understand, and potentially act upon, some of the controversial issues in the news.

Another way of incorporating these applications is to assign projects that require students to undertake a deeper study of one topic from a selection of topics. The goal of such projects is to create a student-centered learning environment, where students are the ones putting forth positions on sociopolitical topics and defending them based on the mathematics they learned. The teacher serves as a guide asking them to examine their sources for conflicts of interest and question discrepancies between academic science and media reports that persist in presenting "both sides" of an issue, when in fact there are no

controversies. A deeper analysis of the science in the larger context of society often reveals strong vested political and financial interests driving beliefs and decisions instead of mathematics and science. Students should be encouraged to think about the trade-offs of human safety vs. corporate bottom lines: who benefits, at whose cost, and why? They need this type of questioning perspective when grappling with complex issues in order to become independent thinkers. Teaching students to connect real-world issues to mathematics and to use it to understand the social and political ramifications so they can act to make changes is education in the fullest sense of the word.

## REFERENCES

Micha Tomkiewicz (2011) *Climate Change: The Fork at the End of Now*, Momentum Press, New York.

Alistair B. Fraser, Bad Science (www.ems. psu.edu/~fraser/Bad/BadGreenhouse.html)

The Intergovernmental Panel on Climate Change, *Climate Change 2001: Working Group I: The Scientific Basis* (www.grida.no/publications/other/ipcc_tar)

Ron Larson and Robert P. Hostetler, *Algebra and Trigonometry*, Houghton Mifflin.

National Research Council of the National Academies, *Surface Temperature Reconstructions for the last 2,000 years* (www.nap.edu/openbook.php?record_id=11676&page=2)

The Pew Forum on Religion and Public Life, *U.S. Religious Landscape Survey—Religious Beliefs and Practices: Diverse and Politically Relevant*, 2008 (religions.pewforum.org/pdf/report2-religious-landscape-study-full.pdf)

Scripps Program, Keeling Curve Lessons, (scrippsco2.ucsd.edu/program_history/keeling_curve_lessons.html)

# CHAPTER THIRTY-ONE

# A Social Justice Data Fair

## BY BETH ALEXANDER AND MICHELLE MUNK

The gym roars with 50 different conversations. Around the walls, and throughout the room, students stand before their projects. Most are pasted on poster board, others on reused pizza boxes. Over the din, the girls describe their work to fellow students, teachers, and a handful of "willing listeners"—volunteer adults (parents, mostly) who have come to find out what the kids have learned.

Along the back wall, the 5th- and 6th-grade classes present their findings from the school waste audit. The 1st- and 2nd-grade students have studied how hard it is to live on the "welfare diet"—the contents of the average food bank recipient's weekly allotment—and have created graphs to show what they, and their families, have eaten that week. The older girls have prepared individual projects. Melanie, 7th grade, wondered if children with autism received the same amount of funding as children with other disorders, and her results are strikingly demonstrated on a bar graph. Mi-sun, 9th grade, used a line graph to demonstrate that Korean students, among all nationalities, are most likely to commit suicide. Andrea, 7th grade, used a scatter plot to investigate the correlation between GDP and carbon dioxide emissions. On a table, the results of the 6th-grade class's collective research is illustrated, with labeled bags of rice offering a three-dimensional representation of rice the world eats every day: China has a huge pile; Iceland just a few grains.

The lights flicker on and off, and the conversations die down. "We're halfway finished," Michelle calls out. "It's time to switch! Those of you who were presenting should now listen, and those who were listening should now present!" The noise picks up again.

The Social Justice Data Fair has been held at the Linden School in Toronto for the past several years. It is an opportunity for our math students to use data management skills to study issues they're concerned about. The fair is an example of how our independent, girl-centered school for students in grades 1 to 12 tries to include topics of social justice in our curriculum. Compared with other Toronto-area private schools, our school is quite culturally diverse, and we enroll students from many nontraditional families. Most of our families support learning math through the lens of social justice. (Parents have commented how much they enjoy having their daughters bring math class conversations to the dinner table.)

David McLimans

We, like all teachers, feel the pressure to cover a packed curriculum in a limited amount of time. In Michelle's case, data management accounts for only a few expectations in her 7th-grade curriculum—what one would expect to cover in a few classes. Preparing students for the data fair takes several weeks, putting the crunch on other important topics. There is increasing evidence, however, that case studies enable students to form meaningful connections with their mathematics and lead to greater learning—especially for girls. The Ontario curriculum, which we follow, also encourages making real-life connections, and this helps justify our program.

The Social Justice Data Fair really "came into its own" by its second year. For the first year, not all classes participated. Some teachers needed to see how it worked before committing to participate. This type of initiative requires all the participating teachers to create meaningful course content beyond their textbooks. It has been exciting to see how each teacher supports students to make those connections in age-appropriate and grade-relevant ways.

## Start by Asking Questions

We teach girls, and girls have historically received messages that they are not good at math. These messages are often self-fulfilling. A social justice pedagogy, in addition to introducing social justice topics into the curriculum, must reduce barriers to learning for all students. It must engage students as agents of change in the world, and encourage them to view their skills—in math and elsewhere—as tools for enacting that change.

The traditional school curriculum takes today's social and environmental conditions mostly for granted. The world just *is*; students are not encouraged to approach it with a curiosity about how it got to be this way. Social justice math, on the other hand, begins with our curiosity about the world and especially about the world's unfairness. Thus a key part of the curriculum at the Linden School is promoting the habit of thoughtful questioning. In math classes, we use the inquiry process to reinforce the idea that knowledge is built from evidence that the students collect themselves. For example, we teach principles of geometry by conducting experiments.

"What do you know about triangles?" Beth asks her grade 5/6 split.

"They have three sides," blurts out one student.

"And three angles," adds another, "and if all the angles are the same, it is an equilateral triangle."

After this, the students find it harder to come up with ideas.

"Are there any other rules that triangles have to follow?" Beth prompts. "For example, do the angles have to be a certain size?"

The students are not sure about this. Slowly, Brianna raises her hand. "It depends on how big the triangle is. If it is bigger, the angles can be bigger."

"Well, let's find out. Today, I want each of you to run an experiment on triangles. Just like you would in science. Think of a question you wonder about triangles, then find an answer. How will you do that?" Beth turns and writes four words on the board: Question, Hypothesis, Observations, Conclusion. "Those are the steps you'll need to follow today."

"Not 'Procedure'?" asks Natane.

"No, not today."

"Yes!" That means less writing. Natane pulls out a piece of paper.

The students are familiar with the scientific method. Through years of in-class investigations and annual science fair projects, we hope they see that the process of science allows anyone to ask a question and to find an answer. Experiments, run by people from all walks of life, formally and informally, form the basis of everything we know about the world around us. It is not a stretch, then, to apply this idea to math. Later, when we prepare for the Social Justice Data Fair, the girls are primed to use math to answer their own questions.

Keisha is constructing different triangles to see what will happen when she adds up the angles.

On her third triangle, she shouts, "Hey, do they always add up to the same thing? I'll try a bigger one." She does, using her ruler to draw a triangle that covers the whole of a new page. Double-checking with her protractor, she writes

David McLimans

down the size of each angle. "Again!" she shouts. The girls around her laugh.

"So," says Beth, "why do you think that happens?"

By the end of the class, Keisha isn't quite sure why her angles always add up to 180 degrees. That's not important, at least not yet. What's important is that she asked a question and found an answer. And the answer, most importantly, didn't come out of the textbook or from the teacher. Social justice education is not just about recognizing problems; it's also about helping students recognize their own power to find answers. They need to trust in their own authority— not just ours.

Something similar is happening in the 7th-grade classroom, where students are learning about percent. Michelle is encouraging students to present different strategies for solving problems. This reinforces the idea that questions can be approached in different ways, that the person doing the inquiry doesn't *uncover* the answer so much as she *constructs* it.

"Who can calculate this percent?" asks Michelle, referring to a problem on the board.

Mahera comes up to the board and writes her solution.

"Is this correct?" Michelle asks the class.

"No, you did the steps in the wrong order," suggests Marla.

"Yes, it's right, but I do it a different way," says Nicole.

"What do others think?" prompts Michelle.

A lively conversation follows, where students discuss the validity and merit of different ways of approaching the problem.

After Michelle gets the students' attention again, she continues the lesson. "Now it's your turn. Come up with your own percent questions."

## Making Meaning Out of Numbers

The Social Justice Data Fair takes place in April, after nearly a full year of practicing inquiry skills in math and in other classes. To prepare for it, the girls start, once again, by asking questions. The 5th/6th-grade class started with a question that had bothered some of the students: Why do the garbage, recycling, and compost get all mixed up in the school containers? After gathering the contents of all of the school's containers, we took the waste up to the roof—the location of our urban school's playground—and, amid some squealing, began the messy, smelly task of sorting and weighing it. The girls were particularly interested in locating "forbidden" trash like disposable Starbucks containers, which are no longer allowed into the school. They also wanted to know how much of the total waste was food packaging, which is discouraged under our Litterless Lunch program. After determining the total masses of categories of waste (like paper and aluminum), as well as measuring the amounts of waste that were sorted incorrectly by the school community, we had an interesting set of primary data. We added this information to the results of our school survey, which gathered information about our population's habits and attitudes when it came to generating and dealing with their garbage. From there, the students used further questions and a variety of graphs to answer them. Which floor generated the most waste? Were we producing more waste than in previous years? Did the location of bins have an effect on how well sorted they were?

The upper grade students, who work independently on their projects, start by selecting an appropriate topic. Some girls find it easy to come up with their topic; they arise out of issues they're passionate about. The 7th-grade class had learned about national and per capita

David McLimans

carbon dioxide emissions when we studied scientific notation, and the school's curriculum has made more and more students aware of the climate crisis, so we had quite a few projects on this topic. Alina and Brigit were both concerned about species extinction, so Alina focused on tigers and Brigit chose elephants. We help students who have trouble choosing something by brainstorming topics as a group, looking in the newspaper for ideas, and describing projects from previous years. The goal is for each girl to be excited about using math to dig deeper into a topic that interests her.

Once the students have selected a topic and have had it approved, the next step is to develop a thoughtful research question. We discuss as a class what makes an investigable question, and then students work to hone their questions into ones that are clear, specific, and can be answered using quantitative data. The students keep in mind that they must produce at least two different types of graphs, and this affects the questions they pose. Sometimes the question development process takes some time, since students must do some preliminary research to find out what kind of data are available. After some research, 7th grader Nicole's concern about homelessness in Toronto turned into two questions: What diseases most commonly affect Toronto's homeless population? And what reasons do homeless people have for not sleeping in a shelter?

The next step is collecting data. Students decide whether it is more appropriate to gather primary or secondary data for their topic and question. For example, Helen, a 7th grader who loves green sea turtles, worried about their declining population. Michelle, her teacher, suggested some online resources, and Helen learned about the age distribution of turtles in the Pacific. Then she analyzed her data to determine what types of graphs would best represent it. From there, she looked for trends and outliers, and was ready to start compiling her work for presentation.

## Fair Day Arrives

The day before the fair, we have a dress rehearsal. The students practice presenting their projects to their classmates, and get feedback. Michelle had planned to act as facilitator for her class, to prompt discussion and to ensure that the space was safe for students to risk asking questions and risk making mistakes. To her delight, no facilitation is necessary: She just sits back and watches math happen.

Heidi stands by her bear-shaped project board. "I learned a lot from this project, but I still want to know if stopping climate change would help the polar bear populations grow again, and I also wonder if polar bears will adapt to having less ice by changing their habits," she concludes her presentation. "Are there any questions?"

A half-dozen hands shoot up. "I really liked your presentation, by the way. I love your board! I was wondering, why did you decide to use a bar graph to represent the populations?" asked Melanie.

"Why did you focus on those regions?" Alina wants to know.

"It's neat how you made the bars different colors," comments Alisa.

"Do you know how they find those numbers?" asks Nita.

Michelle sees students proudly sharing the product of several weeks of hard work,

David McLimans

including background information, graphs, and analysis. She is familiar with the projects, and she has worked with the students to develop good research questions, find data, and use spreadsheet software to create graphs. But what impresses her are the discussions that follow the presentations. Students engage each other on the topics they researched, and on the math they used to analyze their data. They use their data management skills to justify their choices and defend their conclusions. We want students to feel that they have legitimate questions and comments to share about the math, about the thinking that their classmates were sharing.

Before students leave the gym on the afternoon of the fair, they are asked to hand in their "passports"—a brochure in which they've answered a series of questions about the student presentations they attended. The last question is "What did you learn?"

Kayla: "There are more females than males living with low incomes."

Madison: "Black men are more likely to be executed than white men."

Georgia: "8th graders from Singapore have the highest test scores in mathematics."

Meirion: "Green sea turtles are going extinct."

Zaketa: "The average North American purchases 4,000 calories of food each day, but throws 30 percent away."

Nita: "There is too much social justice to fit in this box!" □

---

Names of students have been changed.

# CHAPTER THIRTY-TWO

# The Transnational Capital Auction

## A Game of Survival

### BY BILL BIGELOW

Students are familiar with transnational corporations—Disney, Nike, Sony, Pepsi, Toyota. Corporations are entities with recognizable slogans and logos; they have public faces. Articles in the popular media describe the fortunes of this or that transnational corporation, the behavior of this or that CEO.

But transnational "capital" has no face. It has no jingles, no slogans, no logos. Capital is real, but invisible.

I wrote this simulation game, The Transnational Capital Auction, because I wanted students to grasp some aspects of capital as a force in today's world—to help them see capital as a kind of living being, one with certain needs and inclinations.

The game is a metaphor for the auction that capital holds to determine who in the world will make the most attractive bid for its "services." Students engage this dynamic from the standpoint of Third World elites and simulate a phenomenon that has been called the "race to the bottom," whereby these elites compete against one another to attract capital. The game's "punch line" is an examination of the social and ecological consequences of the auction.

Note: If any of what follows starts to feel complicated, don't worry. The game has been used by lots of teachers, with excellent success. It's more simple than it may appear—and it works. See my article "The Human Lives Behind the Labels" in *Rethinking Globalization* for a description of the activities that I've used before and after the game. (See Resources, page 266.)

## Materials Needed

1. Several desirable candy bars—at least six.
2. Copies of the student handout, "Transnational Capital Auction: Instructions" (page 252)—one for each student in the class.
3. Copies of the "Transnational Capital Auction Credit Sheet" (page 253)—a minimum of seven, enough so that each country group can have at least one.
4. A minimum of 35 copies of the "Bids to Capital" slips (page 254) so that each group has one per round (seven groups, five auction rounds.)

## Suggested Procedure

1. Before the activity, create the auction scoring chart on the board or on an overhead transparency (see page 251).
2. If you have at least 21 students in your class, divide them into seven groups. Ask the groups to form around the classroom, as far away from one another as possible.
3. Distribute copies of "Transnational Capital Auction: Instructions," "Transnational Capital Auction Credit Sheet," and the "Bids to Capital" slips. Read aloud with students "Transnational Capital Auction: Instructions." It should be obvious, but emphasize the distinc-

tion between "Friendly to Capital" credits and game points.

Answer any questions students might have. Review the "Transnational Capital Auction Credit Sheet." Point out that a group earns more credits the friendlier it is to capital. Show them the candy bars and announce that the three groups with the most game points will win all the candy bars.

4. Begin the first round: Tell students to make their bids in each of the categories and to total up their Friendly to Capital credits on the "Bids to Capital" slips. Note that this is really the hardest round because students don't have any way of knowing what the other groups are bidding.

I play "Capital" in the game and wander the classroom as the small groups decide on their bids, urging them to lower the minimum wage, taxes on corporate profits, and the like: "Come on, show that you really want me!"

5. After each group has submitted its bid, write these up on the board or overhead. Award the first game points based on the results—again, 100 game points for the third highest number of Friendly to Capital credits, 50 for the second highest and 25 for the first. (This is explained in the instructions handout.) After this first round has been played, and the points are posted, I scoff at the losers and urge them to get with the program and start making some bids that will attract Capital.

From this point on, for better or worse, the competition to "win," or for candy bars, takes over, and students continue to "race to the bottom" of conditions for their respective countries. Sometimes they even realize what they're doing as they decide on their bids. "This is like 'The Price is Right,' oppression style," I overheard one of my students say one year.

## Quality, Industriousness and Reliability Is What El Salvador Offers You!

**Rosa Martinez produces apparel for U.S. markets on her sewing machine in El Salvador. You can hire her for 57-cents an hour*.**

Rosa is more than just colorful. She and her co-workers are known for their industriousness, reliability and quick learning. They make El Salvador one of the best buys in the C.B.I. In addition, El Salvador has excellent road and sea transportation (including Central America's most modern airport) . . . **and there are no quotas.**

Find out more about **sourcing** in El Salvador. Contact **FUSADES**, the private, non-profit and non-partisan organization promoting social and economic development in El Salvador. Miami telephone: 305/381-8940.

* - Does not include fringe benefits.

**1990**

## Quality, Industriousness and Reliability Is What El Salvador Offers You!

**Rosa Martinez produces apparel for U.S. markets on her sewing machine in El Salvador. You can hire her for 33-cents an hour*.**

Rosa is more than just colorful. She and her co-workers are known for their industriousness, reliablity and quick learning.
They make El Salvador one of the best buys.
In addition, El Salvador has excellent road and sea transportation (including Central America's most modern airport) . . . **and there are no quotas.**

Find out more about **sourcing** in El Salvador. Contact **FUSADES**, the private, non-profit and non-partisan organization promoting social and economic development in El Salvador. Miami telephone: 305/529-2233, Fax: 305/529-9449.

*Does not include fringe benefits/exact amounts may vary slightly depending on daily exchange rates.

**1991**

THESE ADS APPEARED IN TRADE MAGAZINES FOR CLOTHING MANUFACTURERS. NOTE THAT THE HOURLY WAGE DROPPED FROM 57 CENTS AN HOUR IN THE FIRST AD TO 33 CENTS IN THE SECOND.

6. After each round of bids, continue to post the Friendly to Capital credit scores and award game points for that round. Keep a running total of each country team's game points. As Capital, I continue to urge students lower and lower: "Team five, you think I'm going to come to your country if your tax rate is 30 percent? Come on, next round let's get that way down."

7. For the fifth and final round, ask each student to write down their group's last bid, separate from the "Bids to Capital" slips—not just the number of credits, but the actual minimum wage, the child labor laws, etc. For their homework writing assignment, each student needs to know the specific social and environmental conditions created by the auction, in their pretend country.

Award candy bars to the "winners." Distribute the homework assignment, "Transnational Capital Auction Follow-up" (page 255).

8. The next day, we discuss the homework questions on the "Transnational Capital Auction Follow-up" student handout. Some additional questions to raise:

- What does it mean to have no environmental laws in a country? What might Capital do?

- What would be the social effects of such low wages? (You might list these on the board or overhead.) How would families be able to survive? How could a family supplement its income? You might point out that even if a country did have child labor laws, the low wages for adults would put pressure on a family to send its children to work. If students don't point it out, you might also note the relationship between the "race to the bottom" and increased immigration. Of course, there are other factors leading to immigration, such as the dramatic rise in cash crop agriculture which throws peasants off the land. (See, for example, the video *The Business of Hunger,* listed in Resources, page 270.) But it's vital that students begin to see the interconnectedness of global issues.

- Ask students to look at the El Salvador ads from 1990 and 1991 (at left). How are these ads examples of the "race to the bottom"? Who are the ads trying to appeal to?

- With homework question 2 on the handout, I want students to begin to think critically about some of the ideas that are presented as obvious and absolute in much of the media: Is corporate investment always a good thing? You might make two columns on the board or overhead and ask students to list some of the possible benefits of investment and some of the harmful effects of investment and how it is attracted.

Encourage students to ground their answers in their own experiences with the Transnational Capital Auction, as my student Sam did in the following part of his answer to question 2: "Poor countries may need investment, but it is not necessarily a good thing when transnational companies invest capital there. I come to this conclusion by the simple attraction of Capital. Capital is attracted to a place that has less restrictions, on everything. In order to make our country more 'Capital friendly' we were willing to sell out the lives of individuals more and more, step by step. The worse the conditions got, the more Capital was interested."

- Why did you keep driving down conditions in your country? Why didn't you get together and refuse to bid each other down?
- Who benefits and who doesn't benefit from the "race to the bottom" we simulated in class? How could people in various countries get together to stop attacks on their social and environmental conditions?
- What could we do in this country to respond to the global "race to the bottom"?

9. One option is to give students an opportunity to play the Transnational Capital Auction all over again, except this time they could represent labor and environmental activists instead of a country's elite.

It's important that, if possible, students not be left with the sense that the downward leveling they experience in the auction is inexorable. There are things that people can do, and are doing. You might, for example, watch the video *Global Village or Global Pillage* (see Resources, page 270). Students need to see "big pictures" in order to understand seemingly disconnected events in countries around the world. But it's vital that they not feel defeated by this awareness. □

# TRANSNATIONAL CAPITAL AUCTION

## SCORING CHART

| COUNTRY | ROUND 1 | | ROUND 2 | | ROUND 3 | | ROUND 4 | | ROUND 5 | |
|---|---|---|---|---|---|---|---|---|---|---|
| | friendly to capital credits | game points | friendly to capital credits | game points | friendly to capital credits | game points | friendly to capital credits | game points | friendly to capital credits | game points |
| 1 | | | | | | | | | | |
| 2 | | | | | | | | | | |
| 3 | | | | | | | | | | |
| 4 | | | | | | | | | | |
| 5 | | | | | | | | | | |
| 6 | | | | | | | | | | |
| 7 | | | | | | | | | | |

# TRANSNATIONAL CAPITAL AUCTION

## INSTRUCTIONS

You are leaders of a poor country. Each of your countries was either colonized by European countries or dominated by them economically and militarily. You need to attract foreign investment (capital) from transnational corporations for many different reasons.

Of course, not all of your people are poor. Many, including a number of you, are quite wealthy. But your wealth depends largely on making deals with corporations that come to your country. You get various kickbacks, bribes, jobs for members of your families, etc. Some of this is legal; some not. But in order to stay in power you also need to provide jobs for your people, and the owners of capital (companies like Nike, Disney, Coca-Cola, etc.) are the ones who provide thousands of jobs in their factories.

### THE CHALLENGE

The more jobs you can bring into your country, the more legitimacy you have in the eyes of your people. And your government collects taxes from these companies, which help keep your government working and also help you pay back loans to the International Monetary Fund and other foreign-owned banks. The bottom line is this: You badly need these companies to invest capital in your country.

But here's the problem: You must compete with other poor countries that also need capital. Corporations are not stupid, so they let you know that if you want their investment, you must compete with other countries by:

- Keeping workers' wages low.
- Having few laws to regulate conditions of work (overtime, breaks, health and safety conditions, age of workers), or not enforcing the laws that are on the books.
- Having weak environmental laws.
- Making sure that workers can't organize unions; maintaining low taxes on corporate profits, etc.

Basically, companies hold an auction for their investments. The countries who offer the companies the most "freedom" are the ones who get the investment.

### THE GAME

**The goal is to win the game by ending up with the most game points after five auction rounds.** Each country team's goal is to "win" by attracting capital. The team that bids the third highest number of "Friendly to Capital" credits in a round is awarded 100 game points; the team with the second highest number of Capital credits is awarded 50 game points; and the team with the highest number of Capital credits is awarded 25 game points. The other teams get no points for the round. The auction is "silent"—you don't know until the end of each round who has bid what.

Again, Capital will go where the people are "friendliest" to it. However, the "friendlier" you are to Capital, the angrier it may make your own people. For example, Capital wants workers to work for very little and to not worry about environmental laws. But that could start demonstrations or even rebellions, which would not be good for Capital or for you as leaders of your country. That's why the team bidding the highest number of Capital credits does not get the highest number of game points.

### THE LAST RULE

Your team may be the highest (Capital credit) bidder twice and not be penalized. But for each time you are highest bidder more than twice, you lose ten game points—10 the first time, 20 the second, etc. This is a "rebellion penalty."

**Good luck.**

TRANSNATIONAL CAPITAL AUCTION **CREDIT SHEET**

## minimum

### wage/hour credits

| wage/hour | credits |
|---|---|
| $5.00 | 0 |
| $4.75 | 10 |
| $4.50 | 15 |
| $4.25 | 20 |
| $4.00 | 25 |
| $3.75 | 30 |
| $3.50 | 33 |
| $3.25 | 37 |
| $3.00 | 40 |
| $2.75 | 43 |
| $2.50 | 46 |
| $2.25 | 49 |
| $2.00 | 52 |
| $1.75 | 55 |
| $1.50 | 58 |
| $1.25 | 61 |
| $1.00 | 64 |
| $0.85 | 67 |
| $0.75 | 70 |
| $0.65 | 73 |
| $0.55 | 76 |
| $0.45 | 79 |
| $0.35 | 82 |
| $0.30 | 85 |
| $0.25 | 88 |
| $0.20 | 91 |
| $0.15 | 94 |
| $0.10 | 97 |
| $0.05 | 100 |

## child labor

| child labor | credits |
|---|---|
| below age 16 is illegal/enforced | 0 |
| below age 16 is illegal/weakly enforced | 15 |
| below age 16 is illegal/not enforced | 30 |
| below age 14 is illegal/enforced | 50 |
| below age 14 is illegal/weakly enforced | 70 |
| below age 14 is illegal/not enforced | 85 |
| no child labor laws | 100 |

## environmental laws

| environmental laws | credits |
|---|---|
| strict laws/enforced | 0 |
| strict laws/weakly enforced | 15 |
| strict laws/not often enforced | 30 |
| some laws/enforced | 50 |
| some laws/weakly enforced | 70 |
| some laws/not often enforced | 85 |
| almost no environmental laws | 100 |

## worker organizing

| worker organizing | credits |
|---|---|
| unions fully legal/allowed to organize | 0 |
| unions fully legal/some restrictions on right to strike | 15 |
| only government approved unions legal/ some restrictions on right to strike | 30 |
| only government organized unions allowed | 45 |
| unions banned/no right to strike | 60 |
| unions banned/no right to strike/military stationed in factories | 85 |
| unions banned/no right to strike/military stationed in factories/suspected union organizers jailed/military used against strikes | 100 |

## taxation of corporate profits

| rate | credits | rate | credits | rate | credits | rate | credits |
|---|---|---|---|---|---|---|---|
| 75% | 0 | 55% | 20 | 35% | 40 | 15% | 75 |
| 70% | 5 | 50% | 25 | 30% | 50 | 10% | 80 |
| 65% | 10 | 45% | 30 | 25% | 60 | 5% | 90 |
| 60% | 15 | 40% | 35 | 20% | 70 | 0% (no tax) | 100 |

## TRANSNATIONAL CAPITAL AUCTION

### BIDS TO CAPITAL

**country #**

**round #**

minimum wage credits

child labor credits

worker organizing credits

taxation rate credits

environmental laws credits

**TOTAL CREDITS THIS ROUND** _____

## TRANSNATIONAL CAPITAL AUCTION

### BIDS TO CAPITAL

**country #**

**round #**

minimum wage credits

child labor credits

worker organizing credits

taxation rate credits

environmental laws credits

**TOTAL CREDITS THIS ROUND** _____

## TRANSNATIONAL CAPITAL AUCTION

### BIDS TO CAPITAL

**country #**

**round #**

minimum wage credits

child labor credits

worker organizing credits

taxation rate credits

environmental laws credits

**TOTAL CREDITS THIS ROUND** _____

## TRANSNATIONAL CAPITAL AUCTION

### BIDS TO CAPITAL

**country #**

**round #**

minimum wage credits

child labor credits

worker organizing credits

taxation rate credits

environmental laws credits

**TOTAL CREDITS THIS ROUND** _____

# TRANSNATIONAL CAPITAL AUCTION

## FOLLOW-UP

Complete these on a separate sheet of paper.

1.     Look over your auction "bids" for the fifth and final round of the Transnational Capital Auction—on minimum wage, child labor, worker organizing, taxation rates, and environmental laws. If Capital were to accept your "bid" and come to your country, what would be the real human and environmental consequences there? Answer this question in detail.

2.     Based on your experience with the auction, agree and/or disagree with the following statement and back up your answer with evidence: "Poor countries need investment, so it's a good thing when transnational companies invest there."

3.     The global process that we simulated in class is sometimes called "downward leveling" or the "race to the bottom." What, if anything, could people in poor countries do to stop this race to the bottom?

4.     One company used to manufacture all its products in the United States, paying wages that averaged (with benefits) around $16 an hour. The investment director for this company now travels every month to places like Indonesia, El Salvador, and Nicaragua looking for sites to produce his company's products. He says that he would prefer to keep all production in the United States. Based on this simulation and what you know, why do you think this person's company feels forced to send production to countries that have a lot of Friendly to Capital points?

5.     What impact does the race to the bottom have on workers in this country? In what ways might it affect your lives? In answering the question, you might think about the three quotes below.

---

"It is not that foreigners are stealing our jobs,
it is that we are all facing one another's competition."
— **William Baumol,** economist, Princeton University

"Downward leveling is like a cancer that is destroying its host
organism—the earth and its people."
— **Jeremy Brecher and Tim Costello,** *Global Village or Global Pillage*

"Globalization has depressed the wage growth of low-wage workers [in the United States]. It's been a reason for the increasing wage gap between high-wage and low-wage workers."
— **Laura Tyson,** former chair, U.S. Council of Economic Advisers

# SWEATSHOP MATH

## TAKING A CLOSER LOOK AT WHERE KIDS' "STUFF" COMES FROM

BY BOB PETERSON

Students are intrigued about their "stuff" and where their stuff comes from. It's a great entry point for teachers and students to explore various issues, including sweatshops, child labor, and consumerism. In *Rethinking Globalization* (see Resources, page 266) there are many background readings, lesson ideas, and resources to help teachers engage students in thinking deeply about these topics. In addition, *Rethinking Globalization* lists many videos on these subjects that also include some excellent math.

### DO THE MATH

Numerous math opportunities exist when students explore these issues. Among them:

• Tabulate, figure the percentage, and graph which countries students' clothes, bookbags, and shoes come from. Locate all the countries on a wall map using small Post-it notes.

• Using data from *Rethinking Globalization* or from the websites listed below, find the daily, weekly and annual wages of people making common items that people in this country buy. Compare those wages with wages in other countries.

For example: Workers in the Evergreen Factory in the Rio Blanco Industrial Park sew McKids children's clothing, commonly sold at Walmart stores, and make 43 cents an hour. If a worker works a 14-hour shift, how much does he or she make in one day? How much would he or she make in six days? Assuming a six-day work week, with no unpaid sick days or vacation, how much would he or she earn in a year?

### QUESTIONS FOR DEEPER THINKING

• What additional data is necessary to determine how difficult it would be to live on certain wages in a country? How could you find that data?

- Why do companies move their operations to countries where the wages are low?

- When companies move jobs to lower-wage countries, what are the consequences for the workers and communities where the companies originally were located?

- What movements exist to improve the wages and working conditions of people working under these conditions? ∎

---

Excellent websites for collecting data and background information include:

**Rethinking Globalization,** www.rethinkingschools.org/rg

**Institute for Global Labour and Human Rights,** www.globallabourrights.org

**Alliance for Global Justice,** afgj.org

**Clean Clothes Campaign,** www.cleanclothes.org

**Free the Children,** www.freethechildren.org

**Global Exchange,** www.globalexchange.org

**International Labor Organization,** www.ilo.org

# "WORK FASTER!"

## CARTOON TEACHING SUGGESTIONS

Ask students: "What does the cartoonist suggest is the source of wealth?" and "How would one determine a 'fair wage' given various costs of making something?" "What percentage of the cost of the product does the worker get?"

Have students research the average CEO salary and compare it to the average production worker salary. Assume, for the moment, that (a) the CEO gets the full amount that the worker does not get (in this case, $75 to the worker's $25), and (b) the product still costs $100.

### DO THE MATH

Use the data on average CEO and production worker salaries to figure out their relative shares of the $100. Is the cartoon accurate? If not, how should it be adjusted?

Show students the segment of the video *Zoned for Slavery* by the National Labor Committee about the costs of making a Gap shirt in El Salvador. Have students redo the math in the above cartoon and discuss what it means for the people who work.

See "The Organic Goodie Simulation" in *Rethinking Our Classrooms, Vol. 1* (see Resources, page 266) for a great classroom activity that deals with this issue.

## PART 4

David McLimans

# Resources
# for Rethinking
# Mathematics

# Websites

### The Algebra Project
www.algebra.org
Started by civil rights activist Bob Moses, The Algebra Project works for access to higher-level math classes for all people, particularly African Americans.

### The Benjamin Banneker Association, Inc.
www.bannekermath.org
A nonprofit organization of individuals and groups committed to the mathematics education of African American children.

### CorpWatch
www.corpwatch.org
Excellent collection of background articles (and cartoons) on a range of important issues: tobacco, sweatshops, pharmaceuticals, trade agreements, and more.

### Dollars and Sense
www.dollarsandsense.org
Excellent quarterly magazine on economic justice. Check out their books as well. Especially useful for high school-level math resources.

### Ethnomathematics Digital Library
www.ethnomath.org
A collection of over 700 articles, many of them scholarly, on critical math and ethnomathematics by scholars from around the world.

### Fairness & Accuracy in Reporting
www.fair.org
A national media watch group that has offered well-documented criticism of media bias and censorship since 1986. FAIR publishes the excellent, classroom-friendly *Extra!* which is a vital source to get students thinking critically about media coverage of world events.

### Global Exchange
www.globalexchange.org
An organization dedicated to promoting environmental, political, and social justice around the world. In the late 1990s, Global Exchange was perhaps the most important organization drawing attention to Nike's sweatshop abuses. The website details the group's many activities and resources.

### The Harper's Index
www.harpers.org
A monthly statistical snapshot of the world's economic, political, and cultural climate. An excellent source of data for various types of problem-solving.

### The Institute for Global Labour and Human Rights
http://www.globallabourrights.org/
A human rights organization dedicated to the promotion and defense of internationally recognized worker rights in the global economy. Excellent source for reports, photos, and videos of poor working conditions and workers' struggles to improve them.

### Jobs with Justice
www.jwj.org
A national campaign, with local affiliates, to organize support for workers' rights struggles. Their Student Labor Action Project supports student activism around issues of workers' rights as well as social and economic justice.

### National Council of Teachers of Mathematics
www.nctm.org
The principal professional organization of mathematics teachers in the United States. Offers various journals, conferences, and publications.

### National Priorities Project
http://nationalpriorities.org/
An excellent site for articles and facts on a variety of national issues such as the federal budget, the war in Iraq, prescription drugs, the environment, and others.

**Project Seed**

www.projectseed.org

A multifaceted mathematics education program for grades 3–8 that has been around for over 40 years. It uses a Socratic method to teach advanced mathematics to urban youth.

**Radical Math**

www.radicalmath.org

An excellent site developed by math teachers who have assembled hundreds of lessons, charts, graphs, articles, and data sets to help teachers integrate issues of social justice into math classes.

**Rethinking Schools**

www.rethinkingschools.org

Publisher of the book *Rethinking Mathematics*, the quarterly journal *Rethinking Schools,* and over a dozen other books on social justice education.

**SACNAS (Society for the Advancement of Chicanos and Native Americans in Science**

www.sacnas.org

The mission of SACNAS is to encourage Chicano/Latino and Native American students to pursue graduate education and obtain the advanced degrees necessary for science research, leadership, and teaching careers at all levels.

**Teachers for Social Justice** (Chicago)

www.teachersforjustice.org

Website has a calendar of activities and teaching lessons, including several on mathematics.

**Teaching for Change**

www.teachingforchange.org

The Teaching for Change online catalog is a great source for books, videos, and posters for teachers wanting to bring social justice into the classroom.

**TODOS: Mathematics for All**

www.todos-math.org

Advocates for an equitable and high-quality mathematics education for all students, in particular Latino/Hispanic students, by advancing the professional growth and equity awareness of educators.

**United for a Fair Economy**

www.faireconomy.org

An excellent resource for up-to-date data on current social/economic issues. Check out the free "workshops" on Fair Taxes for All, War and the Economy, FTAA for Beginners, The Growing Economic Divide, and others.

**UNICEF (United Nations Children's Fund)**

www.unicef.org

UNICEF produces educational materials and distributes funds to children's programs throughout the world. Their annual report on The State of the World's Children provides useful statistics.

# Math Curriculum and Pedagogy

Bill Carroll, Cheryl Colyer, and Sandie Bulmann, eds. **Math for a Change: 2012 revised edition** (Mathematics Teachers' Association of Chicago and Vicinity, 2012).
Contains 41 situations of injustice that need mathematics to be fully understood. Problems are accompanied by teacher notes and include contexts such as welfare, AIDS, and the savings-and-loans scandals. Grades 7–12.

Ron Eglash, **African Fractals: Modern Computing & Indigenous Design** (Rutgers University Press, 1999).

Helen Featherstone, Sandra Crespo, Lisa M. Jilk, Joy A. Oslund, Amy Noelle Parks, and Marcy B. Wood, **Smarter Together! Collaboration and Equity in the Elementary Math Classroom** (National Council of Teachers of Mathematics, 2011).
This is an excellent, practical guide to using Complex Instruction in mathematics classes. Based on classroom practice and theory, and filled with examples, this text can definitely support teachers in working towards more equitable classrooms.

Marilyn Frankenstein, **Relearning Mathematics: A Different Third R—Radical Maths** (Free Association Books, 1989).
A comprehensive description of how to teach math from a social justice perspective, including chapters on fractions, decimals, percentages, and measurement.

George Gananidis and Molly Gananidis, **How Big Is a Billion?** (Brainy Day, www.brainyday.ca, 2011).

George Gheverghese Joseph, **The Crest of the Peacock: Non-European Roots of Mathematics,** 3rd ed. (Princeton University Press, 2012).

Jacqueline Leonard, **Culturally Specific Pedagogy in the Mathematics Classroom: Strategies for Teachers and Students** (Routledge, 2008).

Robert P. Moses and Charles E. Cobb, **Radical Equations: Math Literacy and Civil Rights** (Beacon Press, 2001).
The authors argue for mathematical literacy as the new "civil right," and they advocate organizing people to demand educational access to powerful mathematics. Moses also tells the story of the voter registration movement in Mississippi during the 1960s which, he argues, holds crucial lessons for today.

National Council of Teachers of Mathematics, **Multicultural and Gender Equity in the Mathematics Classroom: The Gift of Diversity** (NCTM, 1997).
The 1997 NCTM Yearbook. Aimed at classroom teachers, this volume has five sections: issues and perspectives; classroom cultures; curriculum, instruction, and assessment; professional development; and future directions. Mainly short articles, it is generally useful.

Leslie J. Nielson and Michael de Villiers, **Is Democracy Fair? The Mathematics of Voting and Apportionment** (Key Curriculum Press, 1997).
This interesting book grows out of the new South Africa and its freedom from apartheid. In 1994 de Villiers wrote an initial volume (1) to point out that different voting schemes were both mathematical and political choices; (2) to educate people about the different schemes being considered for the new constitution; and (3) to challenge the stereotype that mathematics is only of value to the "hard" sciences. It has excellent teacher notes, is classroom ready, covers many different voting methods, and is very accessible.

Sharan-Jeet Shan and Peter Bailey, **Multiple Factors: Classroom Mathematics for Equality and Justice** (Trentham Books, 1991).
A book from the U.K. about developing anti-racist and pro-justice mathematics teaching. The authors counter Eurocentric approaches to mathematics, examine the history of mathematics, and provide fresh approaches and plenty of examples for teachers, set in explicitly sociopolitical contexts.

David Stinson and Anita Wager, **Teaching Mathematics for Social Justice: Conversations with Educators** (National Council of Teachers of Mathematics, 2012).

David Stocker, **Math That Matters: A Teacher Resource Linking Math and Social Justice** (Canadian Center for Policy, 2006).

Terry Vatter, **Civic Mathematics: Fundamentals in the Context of Social Issues** (Teacher Idea Press, 1996).
A year's worth of middle school mathematics activities using social issues. The material covers the range of basic middle school math areas and is separated into four quarters: Issues of Race and Gender; Poverty and Wealth; The Environment; and Teen Issues. Each section has a range of activities that teachers can adapt for their particular context.

David Whittin and Sandra Wilde, **Read Any Good Math Lately?** Children's Books for Mathematical Learning, K–6 (Heinemann, 1992).
Teaching ideas and bibliographic information useful for integrating literature and mathematics instruction. It does not have a social justice orientation but is helpful nonetheless. Also see *It's the Story That Counts,* by the same authors.

Claudia Zaslavsky, **Africa Counts: Number and Pattern in African Cultures**, 3rd ed. (Lawrence Hill Books, 1999).
Zaslavsky's classic book on the mathematics of Africa covers how numbers and patterns are integral parts of daily and scientific life all over Africa. Includes a thorough debunking of Eurocentric biases and provides rich examples of mathematical recreations, patterns, shapes, counting and numeration systems, geometry, and art—from all around the continent.

Claudia Zaslavsky, **The Multicultural Math Classroom: Bringing in the World** (Heinemann, 1996).
A multicultural perspective for elementary and middle school teachers, including teaching ideas on number systems, counting, calculating, geometry, and data analysis.

Claudia Zaslavsky, **More Math Games and Activities from Around the World** (Chicago Review Press, 2003).
A compilation of 70 games, puzzles, and projects from around the world that help students become interested in and learn math while giving them a greater appreciation for the role of math in many countries. Also see *Math Games and Activities from Around the World* (Chicago Review Press, 1998).

# Math Books with Theoretical/ Academic Perspectives

Leone Burton, **Which Way Social Justice in Mathematics Education?** (Praeger Publishers, 2003).
Provides an international, research-based perspective on the impact of gender, race, ethnicity, culture, and social class in student mathematics learning and achievement. Although the main focus is on gender issues, the ideas touch broader sociocultural issues in mathematics education.

Gilbert Cuevas and Mark Driscoll, eds., **Reaching All Students with Mathematics** (NCTM, 1993).
This NCTM publication contains a lot of information on various programs oriented toward academic achievement and access for students of color and female students. Sections include: A Global View; Changing What Students Learn; Changing How Teachers Teach; Changing How Students Learn.

Brian Greer, Swapna Mukhopadhyay, Arthur B. Powell, and Sharon Nelson-Barber, eds., **Culturally Responsive Mathematics Education** (Routledge, 2009).

Eric Gutstein, **Reading and Writing the World with Mathematics: Toward a Pedagogy for Social Justice** (Routledge, 2006).
A theoretical and practical book about teaching and learning mathematics for social justice, based on a four-year, teacher-research study of the author's own classroom.

Jacqueline Leonard and Danny B. Martin, eds., **The Brilliance of Black Children in Mathematics** (Information Age Publishing, 2013).

Carol Malloy and Laura Brader-Araje, **Challenges in the Mathematics Education of African American Children: Proceedings of the Benjamin Banneker Association Leadership Conference** (NCTM, 1998).
The conference proceedings, keynote address, and research reports from a Banneker Association conference.

Danny Martin, **Mathematics Success and Failure Among African-American Youth** (Erlbaum Associates, 2000).
An excellent study of African American mathematics students in two middle schools, combining sociocultural and cognitive approaches. The author examines issues of student identity and personal agency, community and cultural forces, history, and social context to develop an explanatory theory for why some students succeed while others do not.

Danny Martin, ed., **Mathematics Teaching, Learning, and Liberation in the Lives of Black Children** (Routledge, 2009).
This book critiques and challenges conventional notions about Black children's mathematical understanding and proficiency. In it, the authors collectively and firmly put deficit framings of African American students into the dustbin of history.

David Nelson, George Gheverghese Joseph, and Julian Williams, **Multicultural Mathematics: Teaching Mathematics from a Global Perspective** (Oxford University Press, 1993).
This excellent resource from the U.K. presents a rationale for teaching mathematics with a multicultural and global perspective and provides many examples that teachers can learn from.

Walter G. Secada, series ed., **Changing the Faces of Mathematics** (NCTM, 1999, 2000).
This series covers a wide range of topics, approaches, and programs. The six volumes are: Perspectives on Latinos; Perspectives on Asian Americans and Pacific Islanders; Perspectives on African Americans; Perspectives on

Multiculturalism and Gender Equity; Perspectives on Gender; and Perspectives on Indigenous People of North America. Excellent resources.

Walter G. Secada, Elizabeth Fennema, and Lisa B. Adajian, **New Directions for Equity in Mathematics Education** (Cambridge University Press, 1995).
A collection of articles about mathematics education reform and equity. Themes discussed include social class, race and ethnicity, culture, gender, and language. The articles also deal with teacher deskilling, classroom dynamics, and issues of power.

Ole Skovsmose, **Towards a Philosophy of Critical Mathematics Education** (Kluwer Academic Publishers, 1994).
Don't ask the price of a Kluwer book; only libraries can afford them. But this is the book that lays out a comprehensive and philosophical framework for critical mathematics education. Skovsmose is Danish and writes from that perspective, but the ideas are applicable across the globe.

Na'ilah Suad Nasir and Paul Cobb, eds.), **Improving Access to Mathematics: Diversity and Equity in the Classroom** (Teachers College Press, 2006).

Ole Skovsmose and Brian Greer, eds., **Opening the Cage: Critique and Politics of Mathematics Education** (Sense Publishers, 2012).

Valerie Walkerdine, **Counting Girls Out: Girls and Mathematics**, new ed. (Falmer Press, 1998).
In her classic post-structuralist account of girls in mathematics, Walkerdine argues that dominant ideologies (like patriarchy) play a key role in how we see girls (and boys) in school mathematics. She makes the point that the problem of girls' underachievement in mathematics is not located within girls, but rather in the sociopolitical context of society itself. An important theoretical work that helps us think more deeply about the relationships of gender, inequality, and ideology.

# Curriculum Guides/Resources That Include Math

Wayne Au, Bill Bigelow, and Stan Karp, eds., **Rethinking Our Classrooms, Vol. 1: Teaching for Equity and Justice** (Rethinking Schools, 2007).
A collection of lessons, reflections, poems, and resources for social justice teaching.

Ann and Johnny Baker, **Counting on Small Planets: Activities for Environmental Mathematics** (Heinemann, 1991).
A concise description of how to blend math and environmental education around issues of waste, noise, water, erosion, and deforestation.

Abdul Karim Bangura, **African Mathematics: From Bones to Computers** (Rowman & Littlefield, 2011).

Maurice Bazin and Modesto Tamez, **Math and Science Across Cultures** (New Press, 2002).
A unique collection of activities and lessons on how math is integrated into work and play in different cultures. Lessons include Madagascar Solitaire, The Inca Counting System, and Math in Ancient Egypt.

Bill Bigelow and Norm Diamond, **The Power in Our Hands: A Curriculum on the History of Work and Workers in the United State**s (Monthly Review, 1988).
A widely used curriculum on labor history. Role plays, simulations, first-person readings, and writing activities help students explore issues of work and social change.

Bill Bigelow and Bob Peterson, **Rethinking Globalization: Teaching for Justice in an Unjust World** (Rethinking Schools, 2002).
An excellent resource for teaching strategies that help students make sense of an increasingly

complicated and scary world. Includes role plays, interviews, poetry, stories, background readings, and hands-on teaching tools. Math lessons include Sweatshop Math.

Bill Bigelow, Brenda Harvey, Stan Karp, and Larry Miller, eds., **Rethinking Our Classrooms, Vol. 2: Teaching for Equity and Justice** (Rethinking Schools, 2001).
Extends and deepens many of the themes introduced in *Rethinking Our Classrooms, Vol. 1*, which has sold more than 185,000 copies. Practical, from-the-classroom stories from teachers about how they teach for social justice.

Magi Black, **The No-Nonsense Guide to Water** (Verso Press and the New Internationalist, 2004). Like other books in the No-Nonsense series, this informative handbook is packed with facts, charts, and essays that help teachers and students understand how water has become one of the most important resources in the world and how inequality in access to water is a global crisis. An excellent resource for science and math teachers.

Deborah Menkart, Enid Lee, Margo Okazawa-Rey, eds., **Beyond Heroes and Holidays** (Teaching for Change, 2008). A compilation of teaching and staff development activities that emphasize anti-racist, social justice approaches including ideas on teaching of math.

Tamara Sober Giecek, **Teaching Economics as If People Mattered** (United for a Fair Economy, 2007).
A teacher-friendly guide, full of reproducible graphics and specific teacher lessons on a range of important economic issues including income distribution, wages and salaries of workers and CEOs, globalization and more. High school and above, although some of the charts and activities could be modified for middle school and upper elementary.

Fred Gross, Patrick Morton, and Rachel Poliner, **The Power of Numbers: A Teacher's Guide to Mathematics in a Social Studies Context** (Educators for Social Responsibility, 1993).

Though not explicitly social justice oriented, this book contains some interesting lessons on polling, graphing, and using census figures.

Stephanie Kempf, **Finding Solutions to Hunger: Kids Can Make a Difference** (World Hunger Year, 1997).
A source on hunger for middle and upper school teachers that draws in issues of social studies, math, and politics.

Arthur B. Powell and Marilyn Frankenstein, **Ethnomathematics: Challenging Eurocentrism in Mathematics Education** (SUNY Press, 1997).

Anita Roddick with Brooke Shelby Biggs, **Troubled Water: Saints, Sinners, Truths and Lies About the Global Water Crisis** (Anita Roddick Books, 2004).
This amazing book of eclectic essays, facts, quotations, drawings, and diagrams makes readers look at the pending water crisis from a variety of viewpoints. Excellent for science, social studies, and math. All ages.

Jeremy Seabrook, **The No-Nonsense Guide to World Poverty** (New Internationalist, 2007).
A handbook packed with facts, vignettes, and essays on the roots, scope, and impact of world poverty. An excellent resource for high school social studies and math teachers.

# Additional Books Useful for Mathematics Teaching

Dollars and Sense and United for a Fair Economy, eds., **The Wealth Inequality Reader** (Dollars and Sense, 2004).
A collection of 25 articles that explore wealth inequality, its causes and consequences, and strategies for change. The graphics alone in the opening overview make the book worthwhile for any high school social studies, math, or economics teacher.

Jonathan Teller-Elsberg, James Heintz, and Nancy Folbre, **Field Guide to the U.S. Economy** (New Press, 2007).
An excellent guide to economic life in the United States. Very useful data, graphics, and graphs on a wide range of social issues. Good for upper elementary through college.

John Allen Paulos, **A Mathematician Reads the Newspaper** (Anchor, 1995).
A somewhat cynical look at the mathematical illiteracy of people in the United States, but not necessarily a critical look. Still, a very useful resource for teachers who want to think carefully about promoting critical numeracy.

Dan Smith, **Penguin State of the World Atlas**, 9th ed. (Penguin Books, 2012).
Statistics are given shape and meaning through the visual analysis of data in full-color maps and graphics. Topics include: war, globalization, population growth, human rights, children's rights, health, and more.

Dan Smith, **The Penguin Atlas of War and Peace** (Penguin Books, 2003).
Packed with data analysis and essential political and historical background. Covers the causes of war, the costs of war, and trends in peace, from Afghanistan to Northern Ireland to Kosovo to the Middle East.

Ward L. Kaiser and Denis Wood, **Seeing Through Maps: The Power of Images to Shape Our World View** (ODT, Inc., 2001).
An excellent resource for understanding how maps are created and some of the mathematical issues involved. This book helps us see how maps shape our view of ourselves and others, and examines issues in various world maps (e.g., the Peters and Mercator projections).

# Children's Books

Diana Cohn, **¡Sí Se Puede!/Yes, We Can!** illus. Francisco Delgado (Cinco Puntos Press, 2004). Narrated by a child who tells the story of her mother, a janitor in a Los Angeles high-rise who is out on strike. It portrays dignified people, working hard to make their way into society and fighting for their rights.

Demi, **One Grain of Rice: A Mathematical Folktale** (Scholastic Press, 1997).
A village girl saves her people from starvation by tricking the raja with a mathematical equation. A powerful tale of multiplication and justice.

Judith Ennew, **Exploitation of Children** (Raintree Steck-Vaughn, 1997).
An internationalist perspective on the conditions and types of child exploitation and efforts of people organizing against it. A lot of math, with great potential for use by a creative teacher.

Susan Kuklin, **Iqbal Masih and the Crusaders Against Child Slavery** (Henry Holt and Company, 1998).
An excellent upper elementary biography that sets the short life of Iqbal Masih in the context of the historic struggle against child labor.

David Schwartz, **How Much Is a Million?** illus. Steven Kellogg (Reading Rainbow Books, 1997).
A clear graphic description to help students (and teachers!) conceptualize a million and a billion.

David Smith, **If the World Were a Village: A Book About the World's People**, 2nd ed. illus. Shelagh Armstrong (Kids Can Press, 2011).
A beautifully illustrated explanation of the conditions of the world's people. Makes such topics as language, religion, health, and hunger accessible. A powerful tool for social-justice-minded teachers, particularly teachers of math.

# Sources for Posters, Maps, and Additional Resources

## Maps to Help People Think
www.odt.org
ODT, Inc.
P.O. Box 134, Amherst, MA 01002,
800-736-1293
odtstore@aol.com
Both the Peters Projection Map and the "What's Up? South!" world map challenge people to think differently about the world. The Peters Projection map accurately presents the area of all countries and explains how the commonly used Mercator projection distorts the sizes of continents, making Europe appear much larger than it is. The teaching guide *A New View of the World: Handbook to the Peters Projection World Map,* by Ward L. Kaiser, is also available here.

## Northern Sun Merchandising
www.northernsun.com
2916 E. Lake St., Minneapolis, MN 55406
800-258-8579, fax: 612-729-0149
nsm@scc.net
A distributor of valuable resources on environmental, gay/lesbian, multicultural, and feminist themes. Offers a particularly impressive collection of beautiful, classroom-friendly posters.

## Syracuse Cultural Workers
www.syrculturalworkers.com
P.O. Box 6367, Syracuse, NY 13217
800-949-5139, fax: 800-396-1449
scw@syrculturalworkers.com
A long-time distributor of multicultural and social justice resources, including the Peace Calendar, which should adorn all classrooms.

**Teaching for Change**
www.teachingforchange.org
P.O. Box 73038, Washington, DC 20056
800-763-9131, fax: 202-238-0109
info@teachingforchange.org
Distributor and publisher of quality multi-cultural and social justice teaching materials, such as *Putting the Movement Back into Civil Rights Teaching* and *Beyond Heroes and Holidays*. This organization's online catalog is the single best source for materials to help teachers to explore and teach about social justice issues.

**"Videos with a Global Conscience"** in *Rethinking Globalization: Teaching for Justice in an Unjust World.*
Rethinking Schools
www.rethinkingschools.org/publication/rg
1001 E. Keefe Ave., Milwaukee WI 53212
800-469-6192
Bill Bigelow describes in detail 40 videos, including *The Business of Hunger, Global Village or Global Pillage,* and *Zoned for Slavery,* giving useful teaching ideas. Many of the videos are particularly good for teaching social justice mathematics.

# About the Editors

ERIC (RICO) GUTSTEIN (gutstein@ulc.edu) teaches math education at the University of Illinois at Chicago. He occasionally teaches middle/high school math in Chicago public schools and now works with "Sojo," Chicago's first public high school for social justice, which opened in the fall of 2005. He is the author of *Reading and Writing the World with Mathematics* (Routledge, 2006) and is a co-founder of Teachers for Social Justice (Chicago) and active in the movement against education privatization. He frequently contributes to *Rethinking Schools* magazine.

BOB PETERSON (repmilw@aol.com) is the president of the Milwaukee Teachers' Education Association in Milwaukee, Wis. For many years he taught 5th grade at La Escuela Fratney, a two-way bilingual public school in Milwaukee. He is a founding editor of *Rethinking Schools* magazine and is a frequent writer and speaker. He co-edited (with Bill Bigelow) *Rethinking Globalization: Teaching for Justice in an Unjust World* and *Rethinking Columbus: The Next 500 Years*, and (with Michael Charney) *Transforming Teacher Unions: Fighting for Better Schools and Social Justice*. In 1995 he was selected as Wisconsin Elementary Teacher of the Year.

# Contributors

JULIA M. AGUIRRE is an assistant professor in the education program at the University of Washington Tacoma. Her research focuses on culturally responsive mathematics teaching; equity and social justice in mathematics education; and K-12 mathematics teacher education.

BETH ALEXANDER teaches at the Linden School, an independent girl-centered school in Toronto.

MICHELLE ALLMAN has taught high school mathematics for nearly 15 years, most recently in an alternative high school in Brockton, Mass. She has also worked with districts to support school reform efforts for underserved youth and to develop numeracy-rich classrooms.

SAM E. ANDERSON is a retired New York City Math and Black History Professor, active in the fight against educational genocide by the privatizers of public education, and a founding member of the National Black Education Agenda.

ROBERT BERKMAN teaches mathematics in the New York City area.

BILL BIGELOW is curriculum editor of *Rethinking Schools* magazine and co-directs the Zinn Education Project. He taught high school social studies in Portland, Ore., for almost 30 years and has written and edited many books, including (with Bob Peterson) *Rethinking Globalization: Teaching for Justice in an Unjust World*.

ANDREW BRANTLINGER taught secondary mathematics in Chicago. He teaches at the University of Maryland in the Department of Teaching, Learning, Policy and Leadership.

BRIDGET BREW is a seventh-year teacher. She began her career in Brooklyn, N.Y., before moving to San Francisco. She currently teaches advanced algebra and probability and statistics at June Jordan School for Equity.

JANA DEAN teaches math and science in an interdisciplinary program at Jefferson Middle School in Olympia, Wash. She returned to 6th grade after teaching in the Master in Teaching program at the Evergreen State College and offering professional development in mathematics in southwest Washington school districts.

FLANNERY DENNY teaches 6th- to 8th-grade math at Manhattan Country School, an independent coeducational pre-K to 8th-grade school in New York City with no racial majority, a sliding scale tuition system, and a farm in the Catskills.

INDIGO ESMONDE is an associate professor with the Department of Curriculum, Teaching, and Learning at the Ontario Institute for Studies in Education at the University of Toronto.

SAGE FORBES-GRAY teaches probability and statistics at Sunset Park High School in Brooklyn, N.Y.

MARILYN FRANKENSTEIN teaches quantitative reasoning and arguments at the University of Massachusetts Boston's College of Public and Community Service. She writes about criticalmathematics and ethnomathematics education.

SELENE GONZALEZ-CARILLO has worked as the open space coordinator for Little Village Environmental Justice Organization in Chicago on outreach for Statistics for Action, and works on invasive species management, urban agriculture, and collective action for environmental justice.

MAURA VARLEY GUTIÉRREZ is the director of teaching and learning at Elsie Whitlow Stokes Community Freedom Public Charter School in Washington, D.C. As a graduate student at the University of Arizona and with the Center for the Mathematics Education of Latinas/os (CEMELA: NSF grant ESI-0424983), she had the privilege of working with the young Latinas featured in her article.

SUSAN HERSH was a classroom teacher in elementary and middle schools in the United States and abroad for 10 years. For the past six years, she has been a library media specialist in suburban Milwaukee, Wis., school districts.

KATHRYN HIMMELSTEIN has taught math in New York City and Chicago public schools. She currently teaches algebra at Kenwood Academy High School in Chicago. Racism and Stop and Frisk was taught while she was a mathematics teacher at West Brooklyn Community High School in New York City.

POLLY KELLOGG taught in the Department of Human Relations and Multicultural Education at St. Cloud State University in Minnesota. She is now active with Occupy Minneapolis.

STEPHANIE KEMPF is a teacher and writer in New York City.

SANDRA KINGAN teaches in the Department of Mathematics at CUNY–Brooklyn College in Brooklyn, N.Y.

MICHAEL LANGYEL taught high school mathematics for many years in the Milwaukee Public Schools.

DAVID LEVINE is a co-founder and editor of *Rethinking Schools* magazine. He is a professor of education at Otterbein University in Ohio.

MARTHA MERSON is the project director for Statistics for Action, based at TERC, a not-for-profit committed to science and math education and research. She is a co-author of the EMPower curriculum series (*Extending Mathematical Power*) for nontraditional students enrolled in adult basic education, pre-GED, and transitional courses to college.

RAQUEL MILANI is a doctoral student in mathematics education at the State University of São Paulo, Brazil.

LARRY MILLER is on the school board in Milwaukee, Wis., and is an editor of *Rethinking Schools* magazine. He taught high school for many years.

SWAPNA MUKHOPADHYAY is a mathematics educator who teaches mathematics as a cultural construction. She is a professor in the Graduate School of Education at Portland State University.

MICHELLE MUNK teaches at City View Alternative Senior School, a social justice-focused Grade 7-8 school in Toronto.

ALMANZIA OPEYO is a researcher and data analyst at a large nonprofit in Atlanta dedicated to providing engaging academic and community programs for underserved youth. Previously, she was an adjunct professor at Tufts University and Emerson College, where she taught courses in research methodology and statistical analysis.

LUIS ORTIZ-FRANCO is a professor of mathematics at Chapman University in Orange, Calif. He was a member of the staff of the United Farm Workers where, in collaboration with Cesar Chavez, he used his mathematics skills to secure just and fair labor contracts for farmworkers.

CRYSTAL PROCTOR attended San Francisco public schools and has been a teacher in the San Francisco Unified School District for seven years. She teaches geometry and pre-calculus at June Jordan School for Equity.

JESSICA QUINDEL taught mathematics at Berkeley High School in Berkeley, Calif., from 2004 to 2010 and currently works in the Milwaukee Public Schools Research Department.

ADAM RENNER was an education professor at Bellarmine University before coming to June Jordan School for Equity to become a high school math teacher for the second time in his life. He was a passionate musician, an avid martial artist, and a thoughtful writer. Adam was dedicated to exposing the injustices in society by using math and finding solutions by creating communities. He died in 2010.

LAURIE RUBEL, a former high school mathematics teacher, is an associate professor of secondary education at CUNY-Brooklyn College in Brooklyn, N.Y.

MEGAN STAPLES is an associate professor of mathematics education in the Neag School of Education at the University of Connecticut.

LARRY STEELE teaches at Franklin High School in Seattle.

BEATRIZ FONT STRAWHUN is a clinical mathematics educator in the Department of Education at Trinity University in San Antonio.

WILLIAM F. TATE is the Edward Mallinckrodt Distinguished University Professor in Arts & Sciences at Washington University in St. Louis. He directs the Center for the Study of Regional Competitiveness in Science and Technology and serves as chair of the Department of Education at the university, where he holds academic and research appointments in American culture studies, urban studies, and the Institute for Public Health.

LIZ TREXLER is in her third year of teaching high school math and finishing her M.A. in educational practices, policies, and foundations at University of Colorado at Boulder.

MICHELA TUCHAPESK DA SILVA is a doctoral student in mathematics education at the State University of São Paulo, Brazil.

ERIN E. TURNER is an assistant professor in the Department of Teaching, Learning, & Sociocultural Studies at the University of Arizona. Her research focuses on issues of equity and social justice in mathematics education, and on mathematics teaching and learning in culturally and linguistically diverse settings.

CLAUDIA ZASLAVSKY (1917–2006) was a noted teacher of mathematics and author of 13 books and many articles dealing with multicultural perspectives and equity issues in mathematics education. Her last book was for children: *More Math Games & Activities from Around the World* (Chicago Review Press, 2003).

MARIA DEL ROSARIO ZAVALA is an assistant professor of elementary mathematics and bilingual education at San Francisco State University, with experience teaching in multiple settings across K-12.

# Index

Page numbers in *italics* indicate illustrations

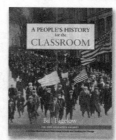

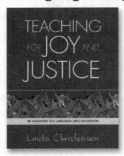

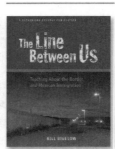

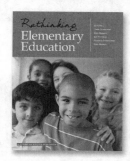
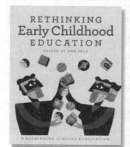